Learn

Eureka Math™
Grade K
Module 4

Published by Great Minds®.

Printed in the U.S.A.
This book may be purchased from the publisher at eureka-math.org.
10 9 8 7 6 5 4 3 2 1

v1.0 PAH

ISBN 978-1-64054-078-1

GK-M4-L-05.2018

Learn ◆ Practice ◆ Succeed

Eureka Math™ student materials for *A Story of Units*® (K–5) are available in the *Learn, Practice, Succeed* trio. This series supports differentiation and remediation while keeping student materials organized and accessible. Educators will find that the *Learn, Practice,* and *Succeed* series also offers coherent—and therefore, more effective—resources for Response to Intervention (RTI), extra practice, and summer learning.

Learn

Eureka Math Learn serves as a student's in-class companion where they show their thinking, share what they know, and watch their knowledge build every day. *Learn* assembles the daily classwork—Application Problems, Exit Tickets, Problem Sets, templates—in an easily stored and navigated volume.

Practice

Each *Eureka Math* lesson begins with a series of energetic, joyous fluency activities, including those found in *Eureka Math Practice*. Students who are fluent in their math facts can master more material more deeply. With *Practice*, students build competence in newly acquired skills and reinforce previous learning in preparation for the next lesson.

Together, *Learn* and *Practice* provide all the print materials students will use for their core math instruction.

Succeed

Eureka Math Succeed enables students to work individually toward mastery. These additional problem sets align lesson by lesson with classroom instruction, making them ideal for use as homework or extra practice. Each problem set is accompanied by a Homework Helper, a set of worked examples that illustrate how to solve similar problems.

Teachers and tutors can use *Succeed* books from prior grade levels as curriculum-consistent tools for filling gaps in foundational knowledge. Students will thrive and progress more quickly as familiar models facilitate connections to their current grade-level content.

Students, families, and educators:

Thank you for being part of the *Eureka Math*™ community, where we celebrate the joy, wonder, and thrill of mathematics.

In the *Eureka Math* classroom, new learning is activated through rich experiences and dialogue. The *Learn* book puts in each student's hands the prompts and problem sequences they need to express and consolidate their learning in class.

What is in the Learn *book?*

Application Problems: Problem solving in a real-world context is a daily part of *Eureka Math*. Students build confidence and perseverance as they apply their knowledge in new and varied situations. The curriculum encourages students to use the RDW process—Read the problem, Draw to make sense of the problem, and Write an equation and a solution. Teachers facilitate as students share their work and explain their solution strategies to one another.

Problem Sets: A carefully sequenced Problem Set provides an in-class opportunity for independent work, with multiple entry points for differentiation. Teachers can use the Preparation and Customization process to select "Must Do" problems for each student. Some students will complete more problems than others; what is important is that all students have a 10-minute period to immediately exercise what they've learned, with light support from their teacher.

Students bring the Problem Set with them to the culminating point of each lesson: the Student Debrief. Here, students reflect with their peers and their teacher, articulating and consolidating what they wondered, noticed, and learned that day.

Exit Tickets: Students show their teacher what they know through their work on the daily Exit Ticket. This check for understanding provides the teacher with valuable real-time evidence of the efficacy of that day's instruction, giving critical insight into where to focus next.

Templates: From time to time, the Application Problem, Problem Set, or other classroom activity requires that students have their own copy of a picture, reusable model, or data set. Each of these templates is provided with the first lesson that requires it.

Where can I learn more about Eureka Math *resources?*

The Great Minds® team is committed to supporting students, families, and educators with an ever-growing library of resources, available at eureka-math.org. The website also offers inspiring stories of success in the *Eureka Math* community. Share your insights and accomplishments with fellow users by becoming a *Eureka Math* Champion.

Best wishes for a year filled with aha moments!

Jill Diniz

Jill Diniz
Director of Mathematics
Great Minds

The Read–Draw–Write Process

The *Eureka Math* curriculum supports students as they problem-solve by using a simple, repeatable process introduced by the teacher. The Read–Draw–Write (RDW) process calls for students to

1. Read the problem.
2. Draw and label.
3. Write an equation.
4. Write a word sentence (statement).

Educators are encouraged to scaffold the process by interjecting questions such as

- What do you see?
- Can you draw something?
- What conclusions can you make from your drawing?

The more students participate in reasoning through problems with this systematic, open approach, the more they internalize the thought process and apply it instinctively for years to come.

Contents

Module 4: Number Pairs, Addition and Subtraction to 10

Topic A: Compositions and Decompositions of 2, 3, 4, and 5

Lesson 1 1
Lesson 2 7
Lesson 3 11
Lesson 4 15
Lesson 5 19
Lesson 6 23

Topic B: Decompositions of 6, 7, and 8 into Number Pairs

Lesson 7 27
Lesson 8 31
Lesson 9 35
Lesson 10 39
Lesson 11 43
Lesson 12 47

Topic C: Addition with Totals of 6, 7, and 8

Lesson 13 53
Lesson 14 57
Lesson 15 63
Lesson 16 67
Lesson 17 71
Lesson 18 77

Topic D: Subtraction from Numbers to 8

Lesson 19 81
Lesson 20 85
Lesson 21 89
Lesson 22 93
Lesson 23 97
Lesson 24 101

Topic E: Decompositions of 9 and 10 into Number Pairs

Lesson 25 105

Lesson 26 109

Lesson 27 113

Lesson 28 117

Topic F: Addition with Totals of 9 and 10

Lesson 29 121

Lesson 30 125

Lesson 31 131

Lesson 32 137

Topic G: Subtraction from 9 and 10

Lesson 33 143

Lesson 34 149

Lesson 35 153

Lesson 36 157

Topic H: Patterns with Adding 0 and 1 and Making 10

Lesson 37 161

Lesson 38 165

Lesson 39 169

Lesson 40 173

Julia found 3 seashells.

Megan found 2 seashells.

How many did they find in all?

Draw

Draw the seashells the girls found. Tell your partner how many seashells the girls found.

Name ____________________ Date __________

Draw the light butterflies in the number bond. Then, draw the dark butterflies. Show what happens when you put the butterflies together.

Name ______________________ Date ____________

How many ? ☐ How many ? ☐

Draw to show how to take apart the group of cats to show 2 groups, the ones sleeping and the ones awake.

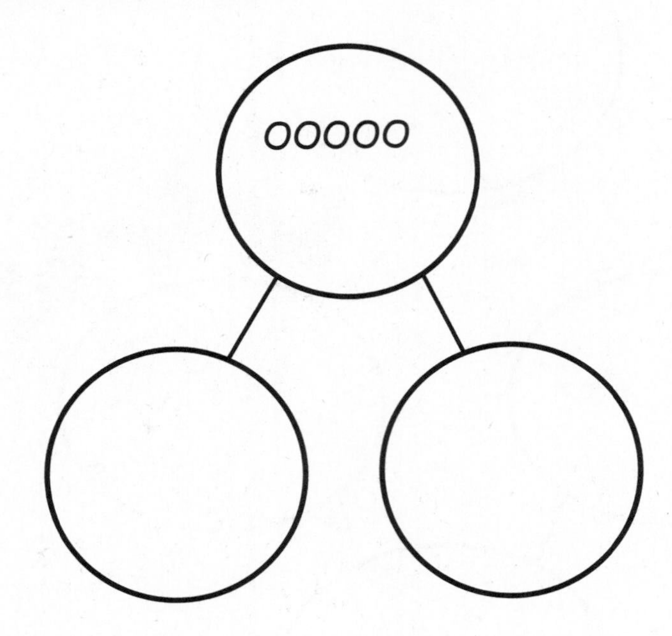

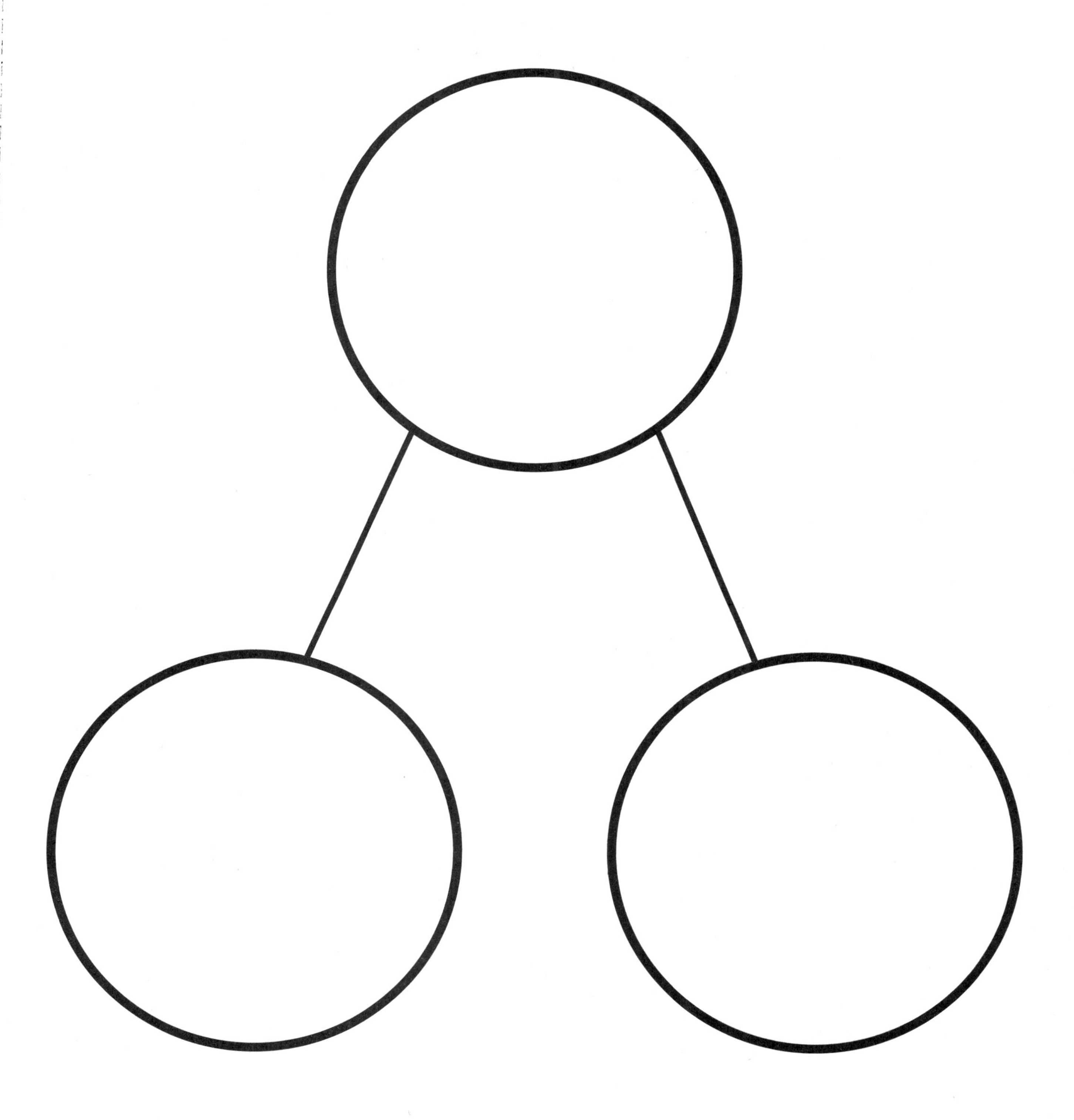

number bond

Together, Margaret and Caleb have 5 pennies.

Draw

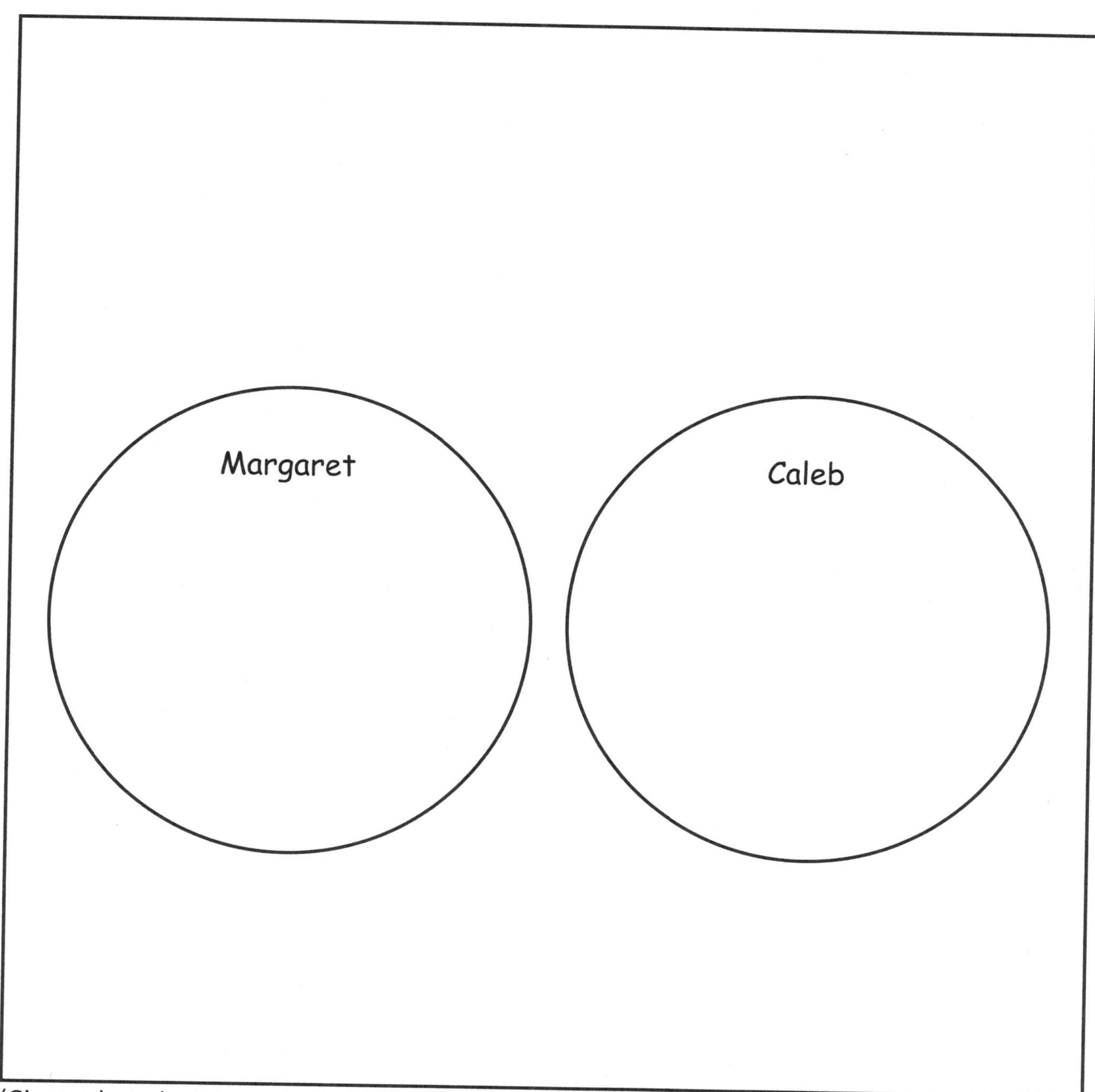

(Give each student 5 pennies.) Take your 5 pennies. Put some pennies in Margaret's circle. Put the other pennies in Caleb's circle. Tell your friend how many pennies Margaret and Caleb each have. Could you take apart the pennies in a different way?

Name ______________________ Date ____________

The squares below represent a cube stick. Color the squares to match the rabbits. 4 squares gray. 1 square black. Draw the squares in the number bond.

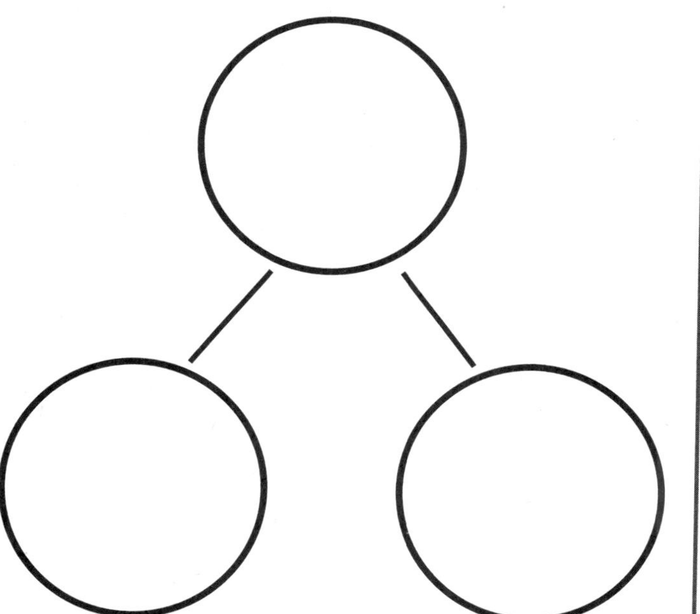

Show the parts of the number bond on your fingers. Color the fingers you used.

☐ rabbits and ☐ rabbit make ☐ rabbits.

Chris has 3 baseball cards.

Katharine has 2 baseball cards.

How many cards do they have together?

Draw

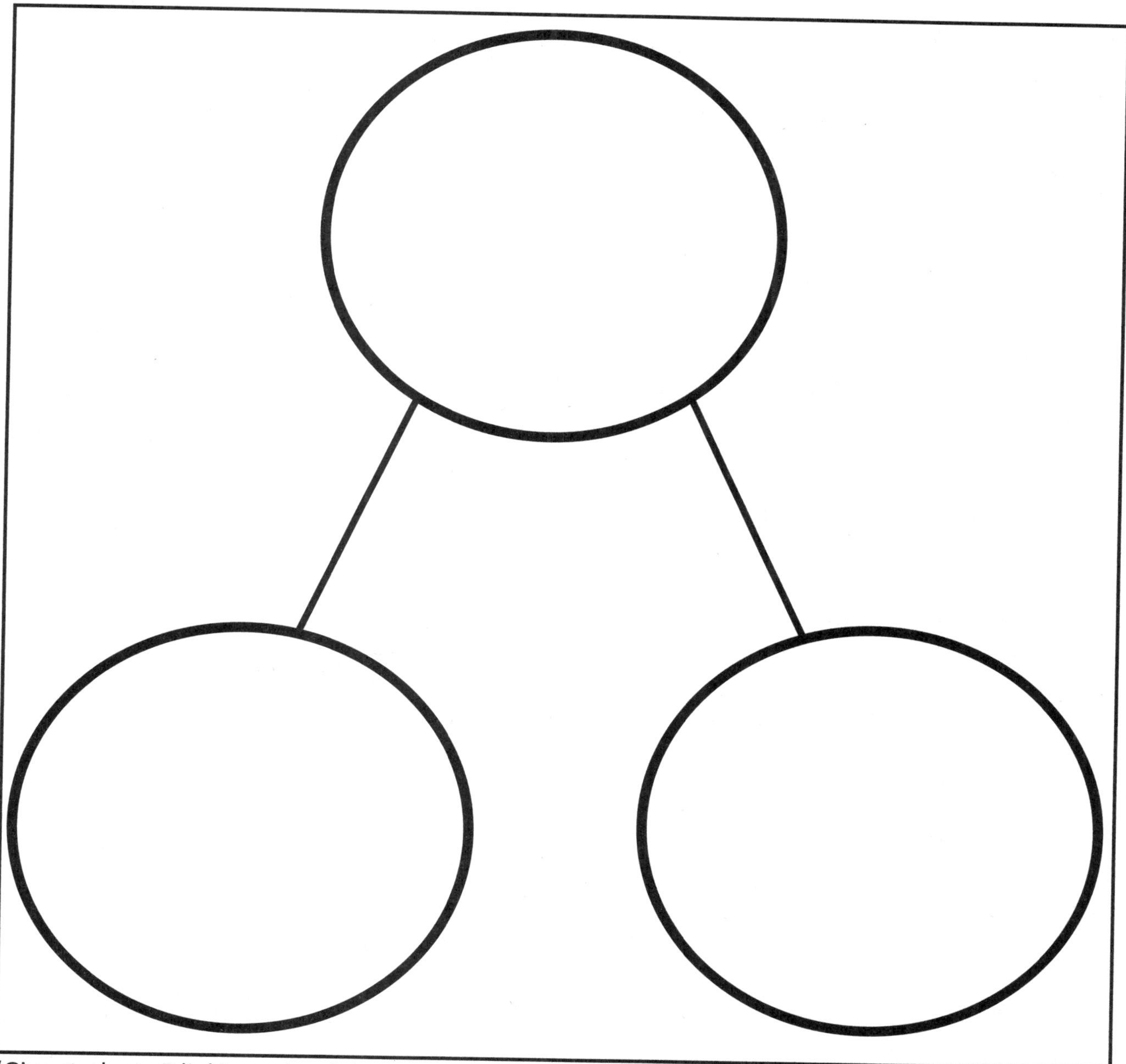

(Give students 5 linking cubes.) Use your linking cubes to show the story. Draw a picture and make a number bond about your story. Talk about your work with a partner.

Name ______________________________ Date ______________

Draw the shapes and write the numbers to complete the number bonds.

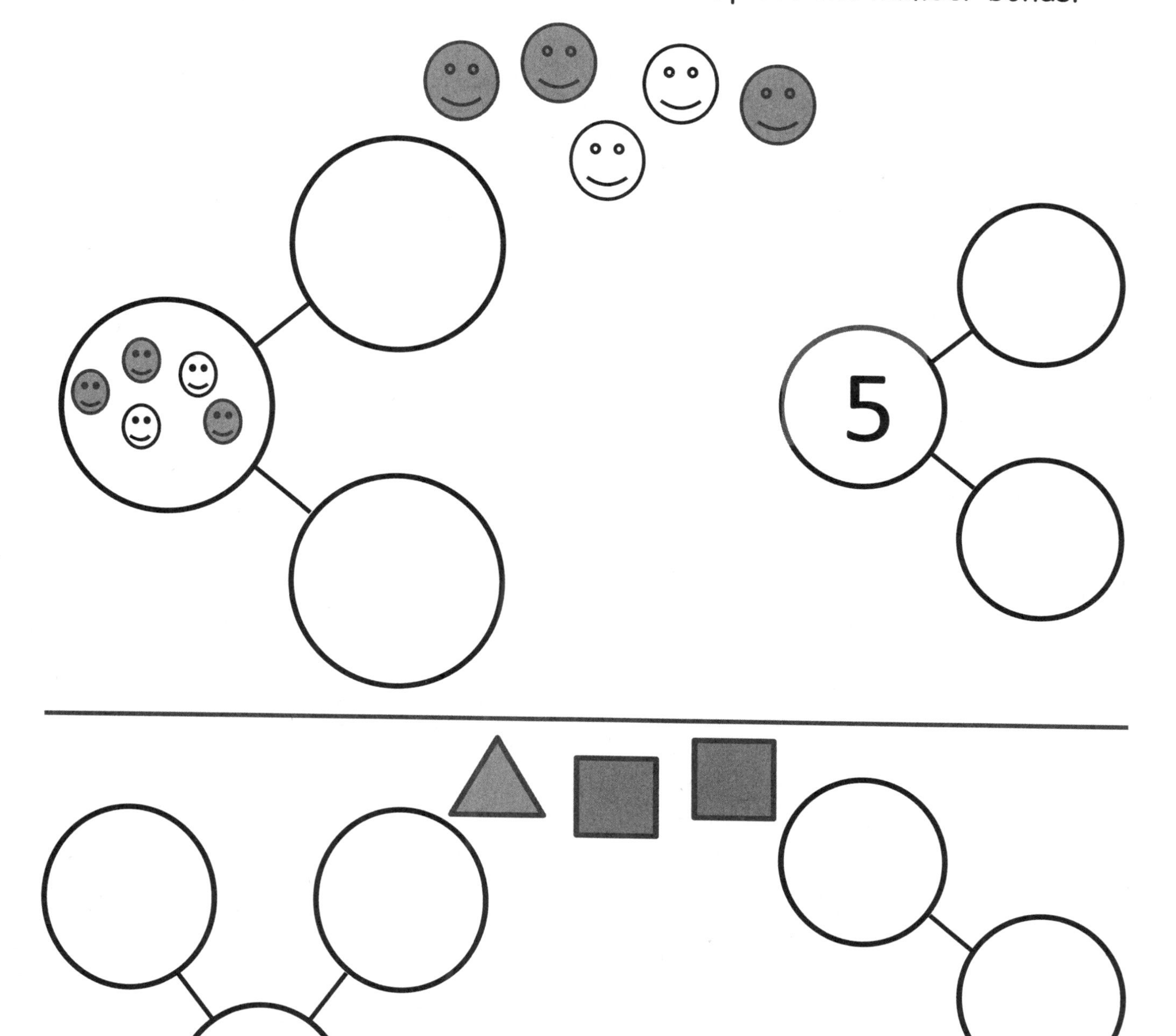

Write numbers to complete the number bond. Put the dogs in one part and the balls in the other part.

Look at the picture. Tell a story about the birds going home to your neighbor. Draw a number bond, and write numbers that match your story.

Anthony had 5 bananas.

Draw

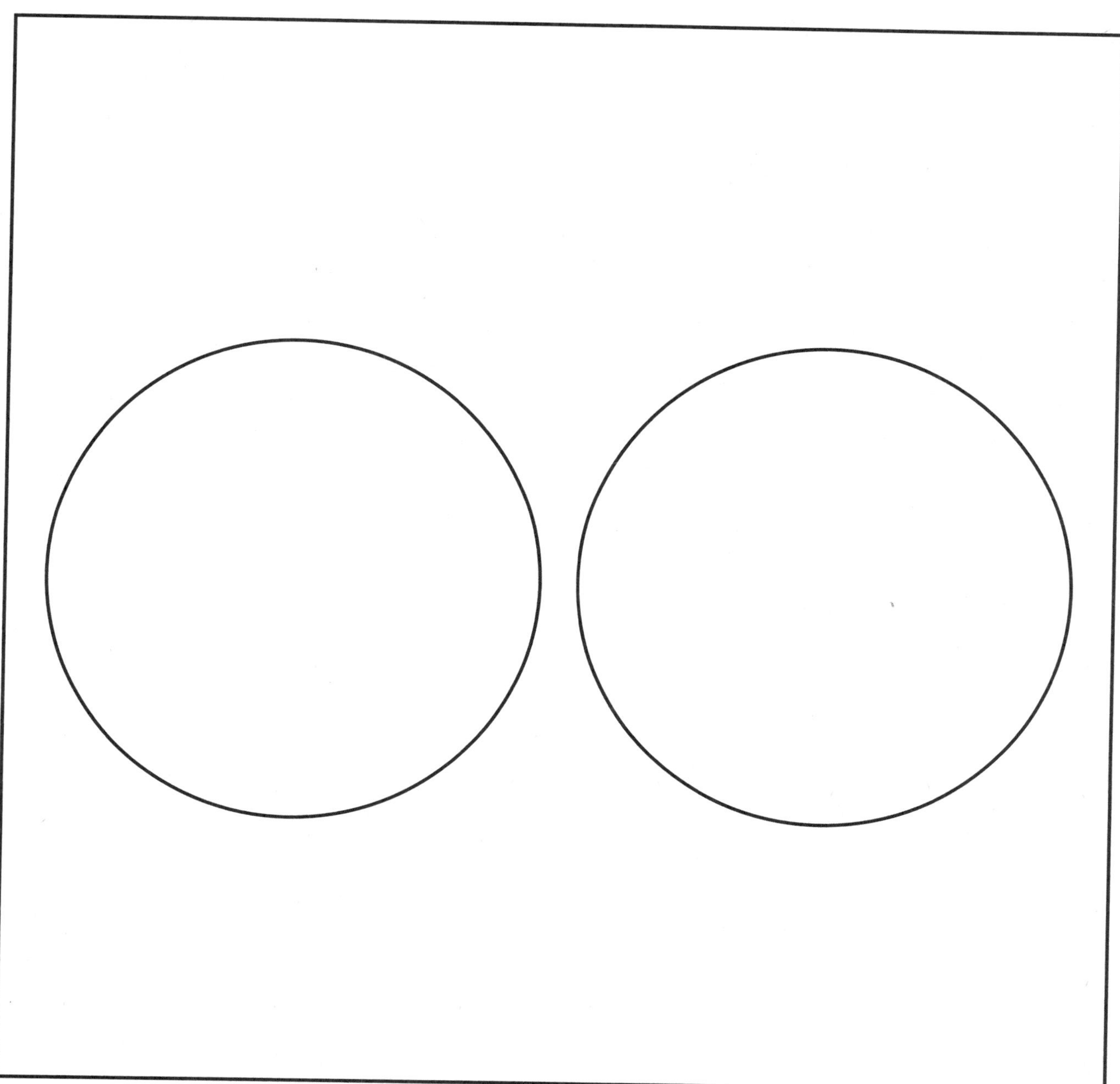

(Give students a small piece of clay.) Make the 5 bananas with your clay. Anthony wanted to share the bananas with a friend. Put the bananas on the circles to show one way he could share the bananas with his friend. Draw a number bond to show how he shared his 5 bananas.

Name ______________________ Date ____________

Draw and write the numbers to complete the number bonds.

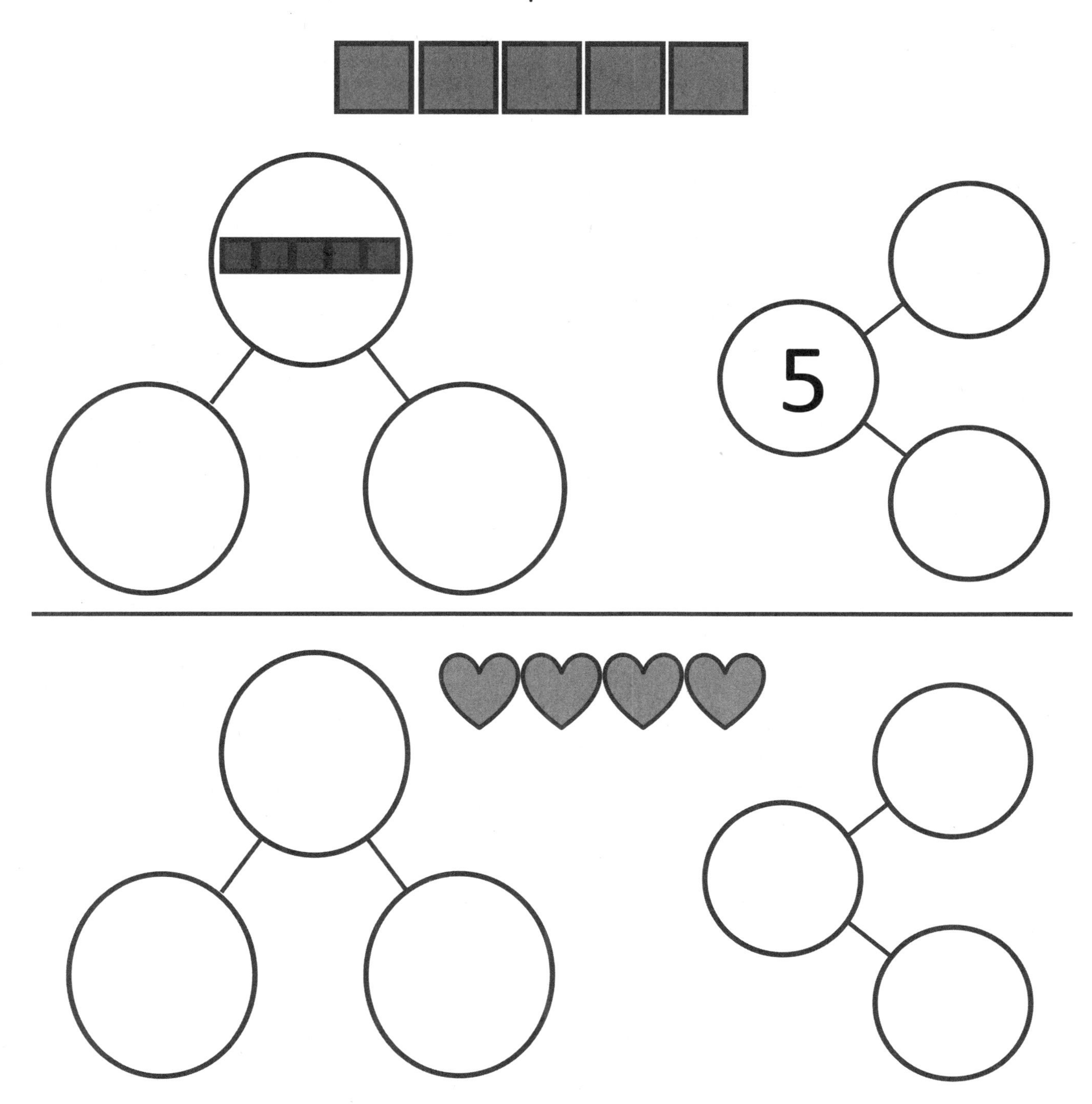

Look at the picture. Tell your neighbor a story about the dogs standing and sitting. Draw a number bond, and write numbers that match your story.

A puppy had 5 bones.

He buried some in the yard and put some by his dish.

Draw his bones.

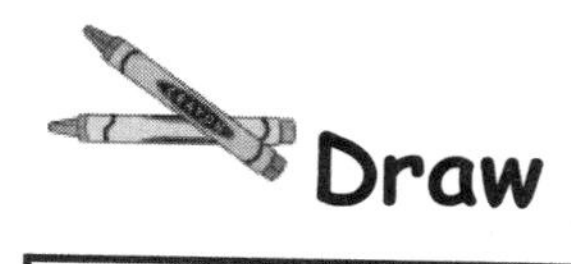

Draw

Compare your picture with a friend's. Did you make your pictures the same? Tell your friend how your pictures are alike. Tell your friend how your pictures are different.

Name ______________________________ Date ______________

Write numbers to fill in the number bonds.

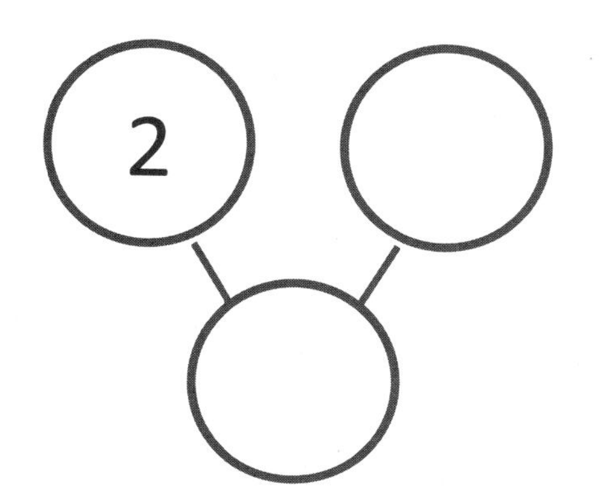

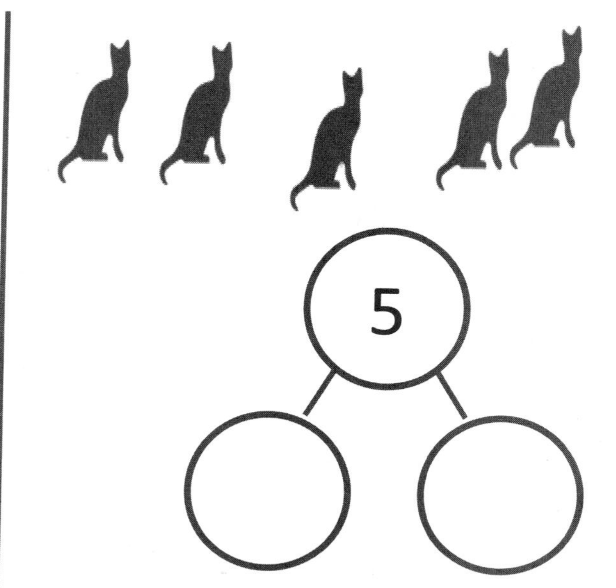

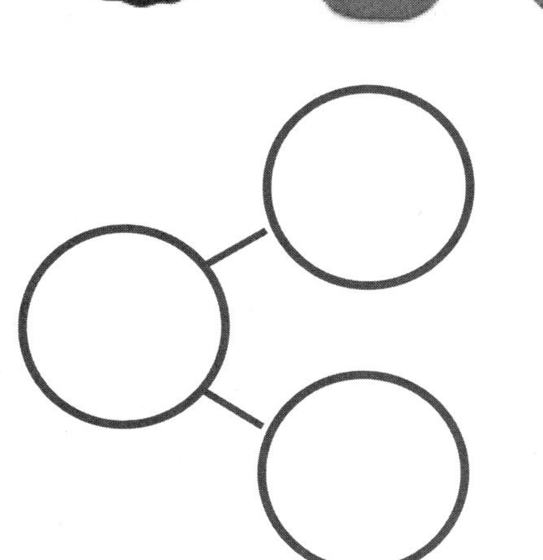

Play a game called Snap with a friend.

Draw

(Give students a 5-stick of linking cubes.) Take turns with a partner. Hold your 5-stick behind your back. When your partner says, “Snap!” break it into two parts. Show one of the parts while keeping the other behind your back. Can your partner guess the other part? Show the missing piece. Draw a number bond to show the two parts you made.

Name ______________________________ Date ______________

Fill in the number bond. Tell a story about the birds to your friend.

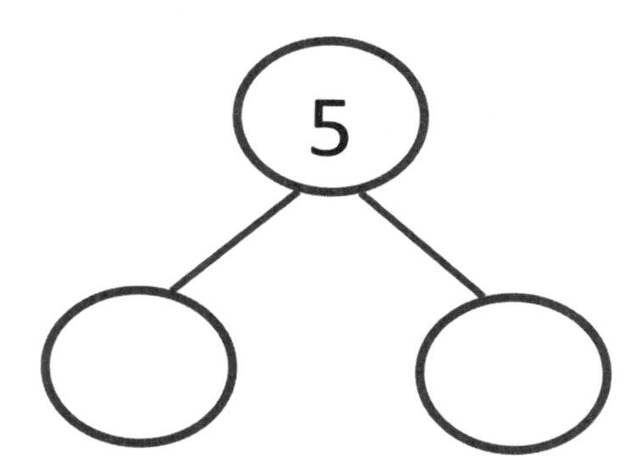

Tell a story that matches the number bond. Draw pictures that match your story.

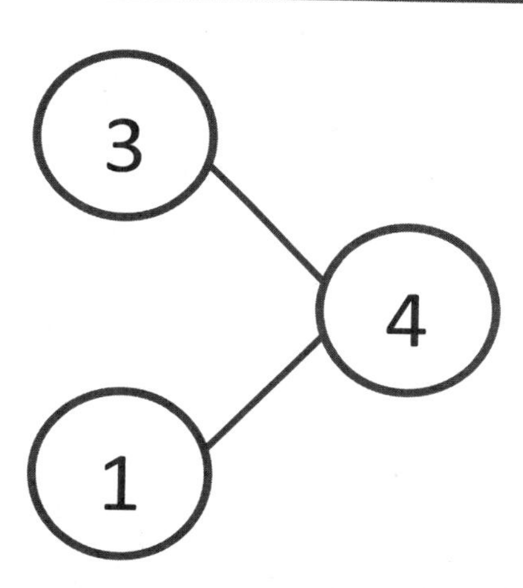

Tell a story. Draw pictures and a number bond that match your story.

The squares below represent cube sticks. Draw a line to match the number bond to the cube stick.

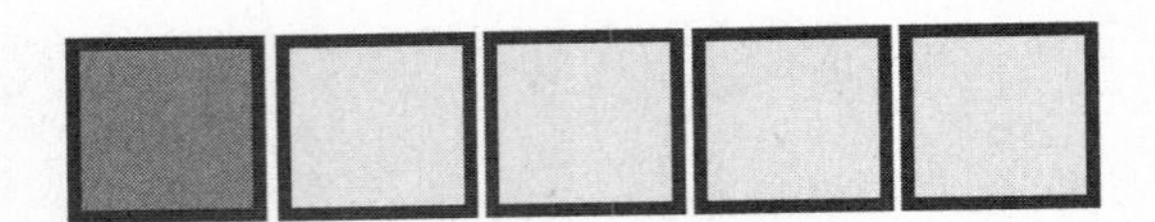

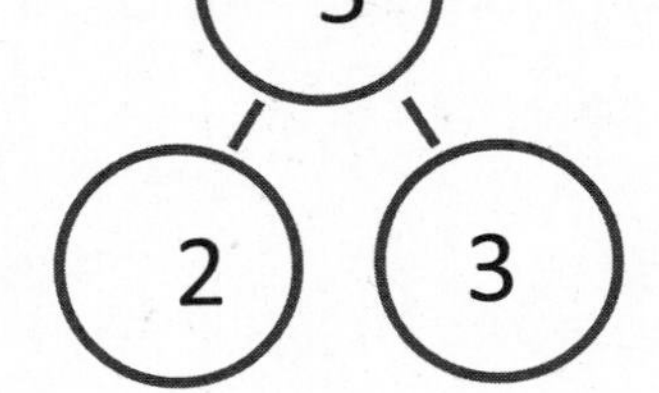

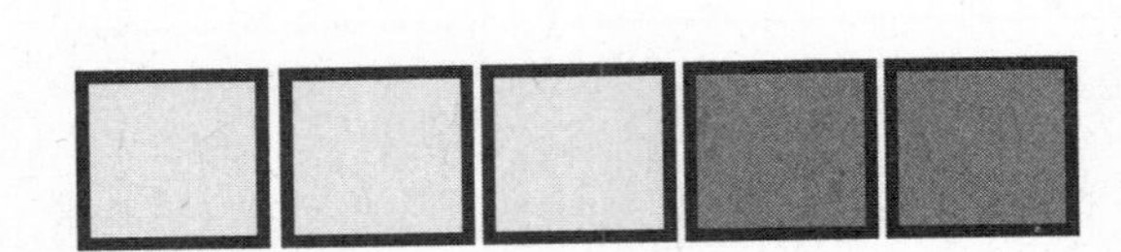

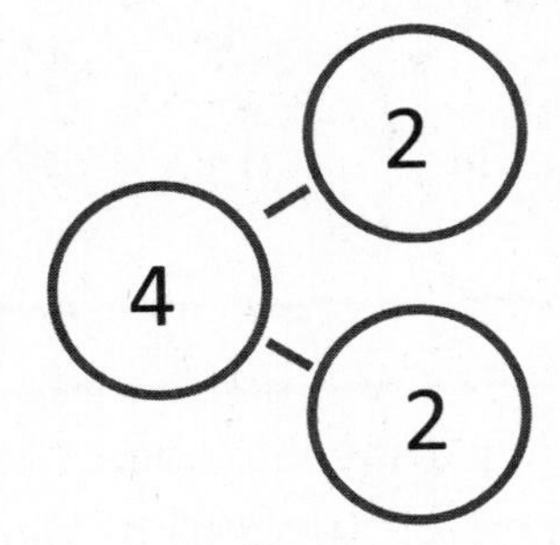

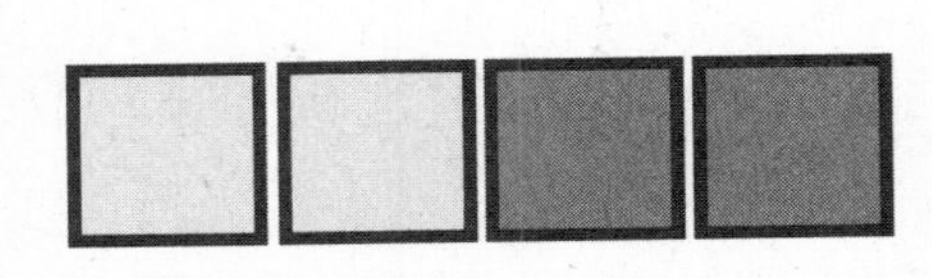

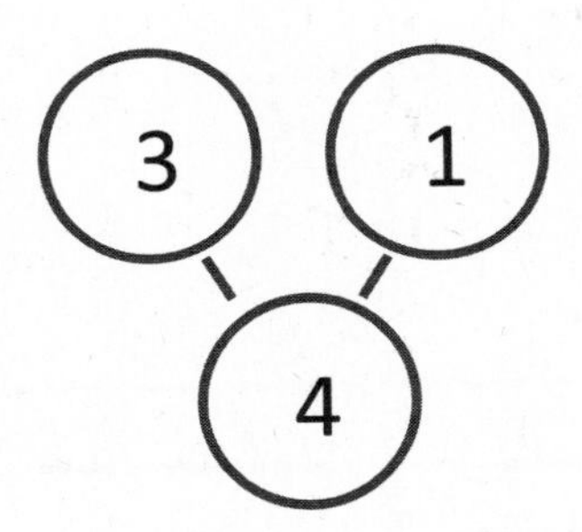

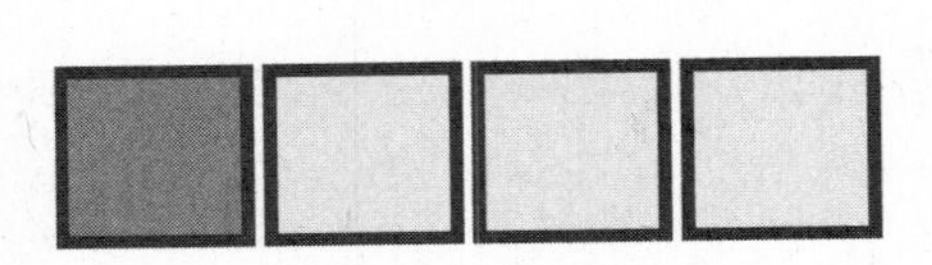

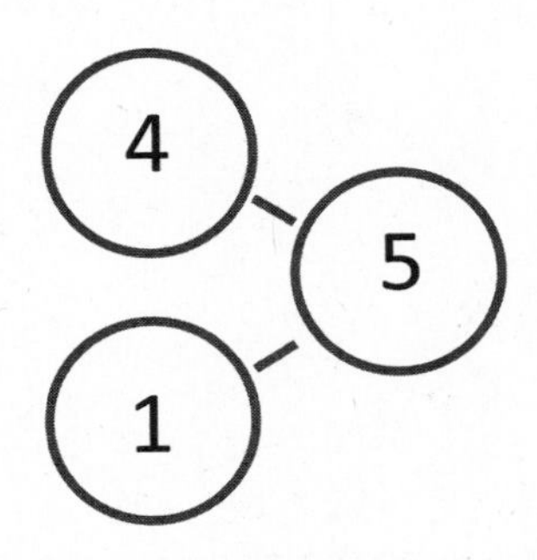

Draw a number bond to show how many claps.

Draw

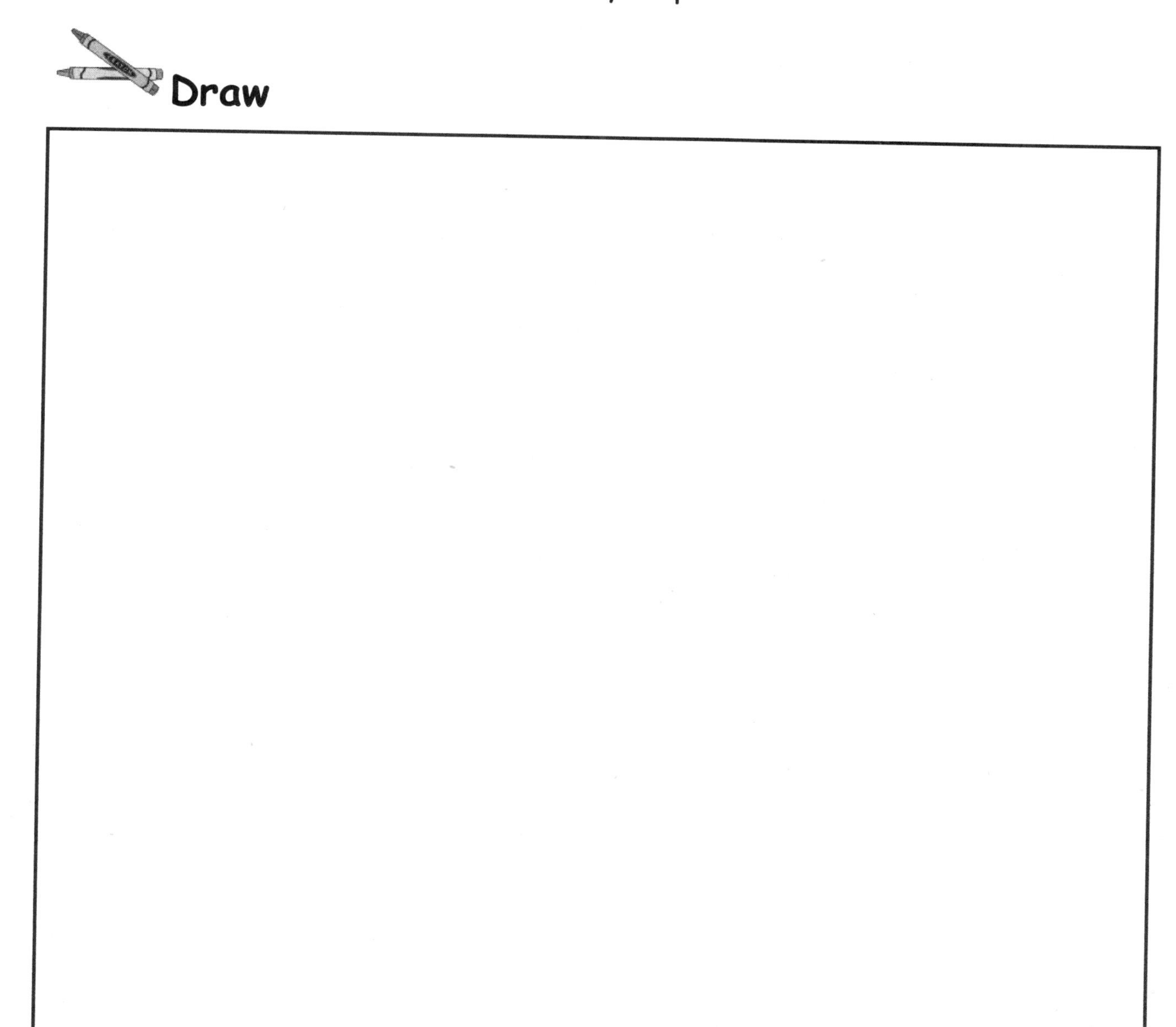

Close your eyes, and count each time I clap. (Clap 5 times; pause, and then clap 1 more time.) Open your eyes. How many claps did you hear? (Allow time for students to answer.) Let's do it 1 more time. (Repeat.) How many claps did you hear? What is 1 more than 5? (Repeat the exercise several times, using claps and instrument sound parts of 4 and 2, 3 and 3, 2 and 4, and 1 and 5.) Now, try the game with a partner. Take turns clapping different **number partners** for 6.

Name ______________________ Date __________

Look at the birds. Make 2 different number bonds. Tell a friend about the numbers you put in one of the bonds.

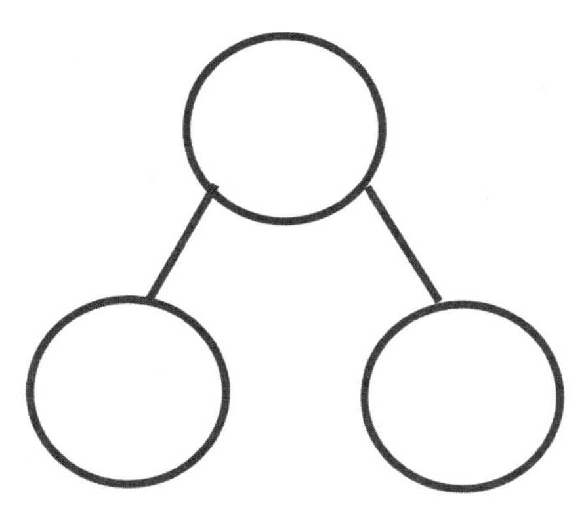 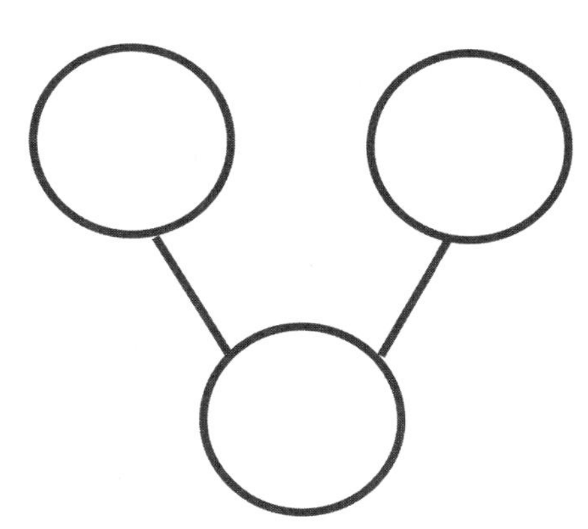

Color some squares green and the rest yellow. Write numbers in the bonds to match the colors of your squares.

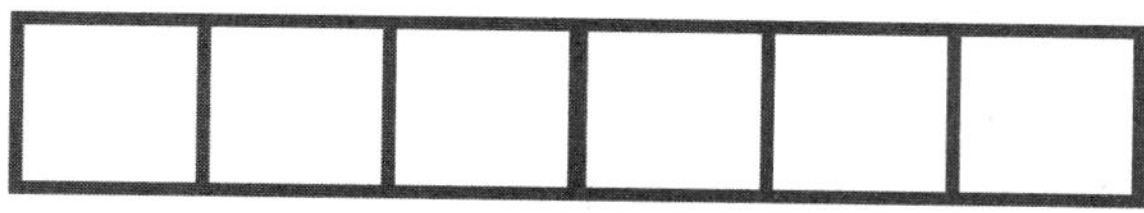

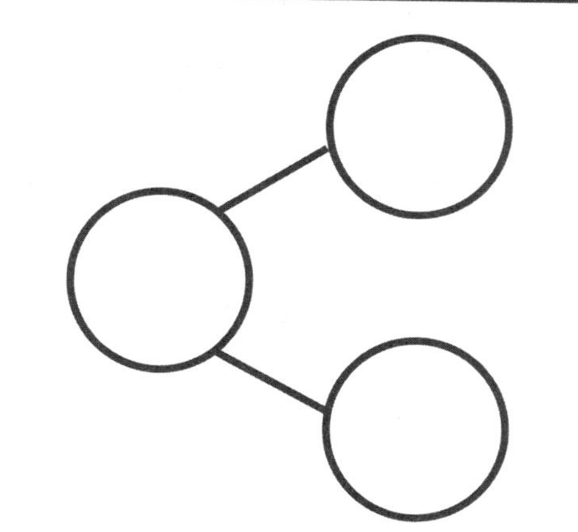

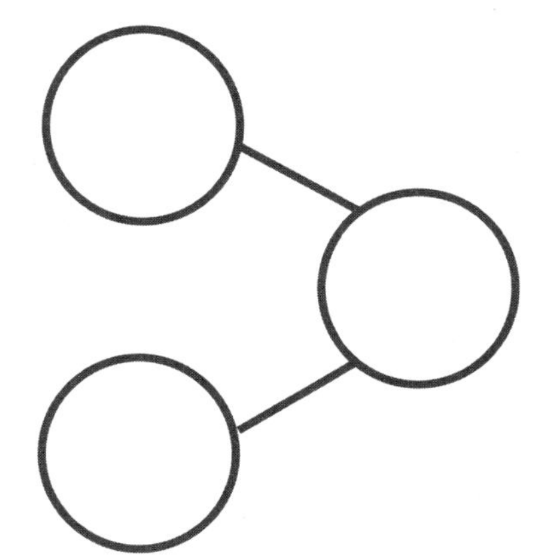

Ming has 5 raisins.

Dan has 2 raisins.

How many raisins are there in all?

Draw

(Give students a ball of clay.) Use the clay to show Ming's and Dan's raisins. Put Ming's raisins into a 5-group. Put Dan's raisins in a row under Ming's raisins. Draw a number bond to show Ming's and Dan's raisins.

Name ______________________ Date ____________

Tell a story about the shapes. Complete the number bond.

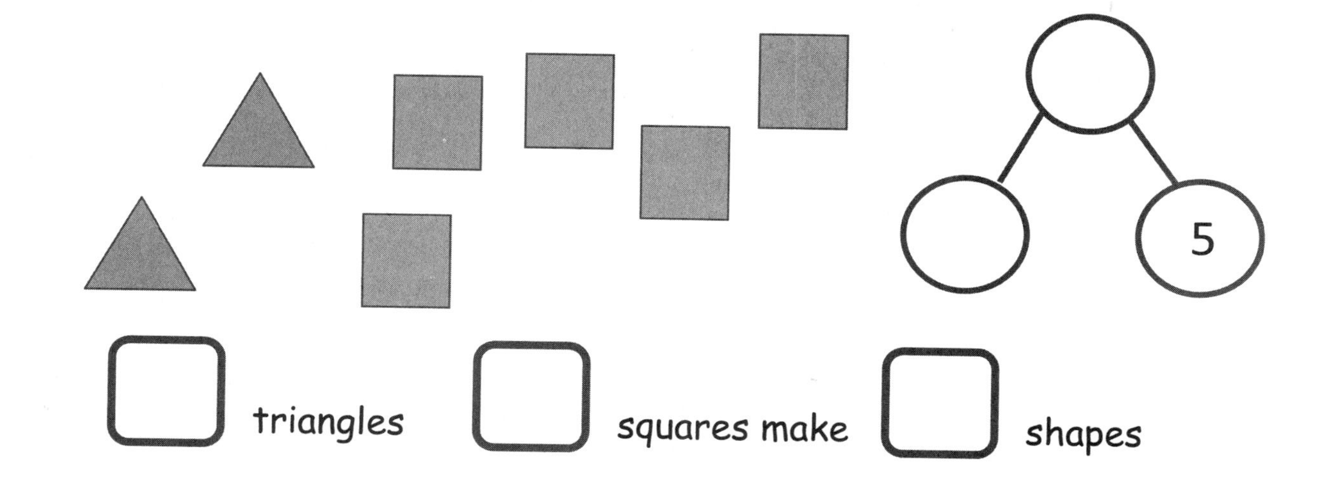
5
triangles
squares make
shapes

The squares below represent cube sticks. Color the cube stick to match the number bond.

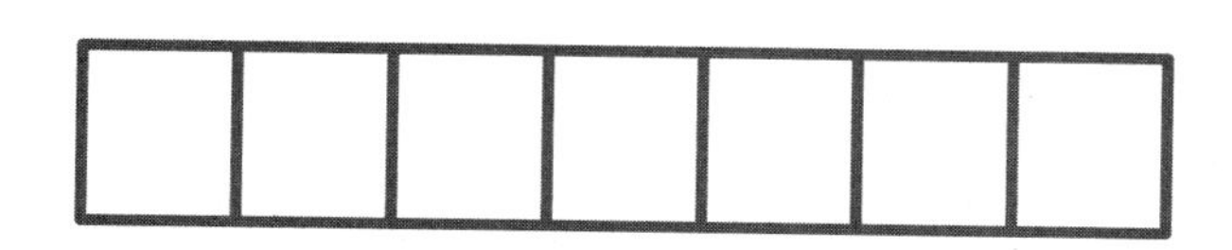

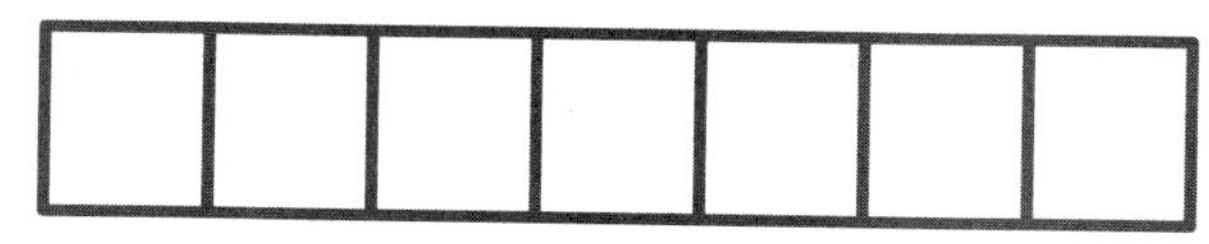

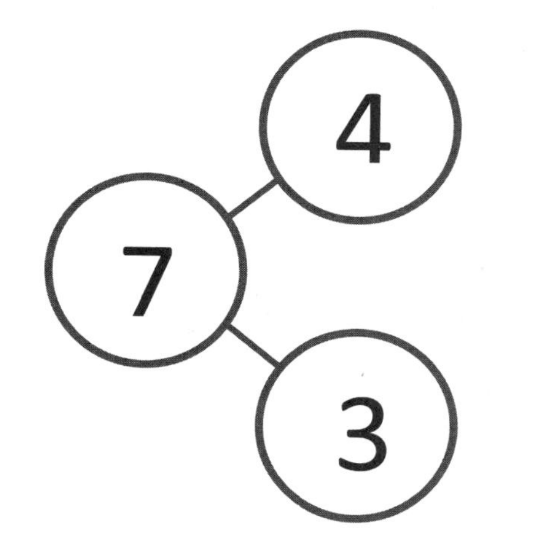
4
7
3

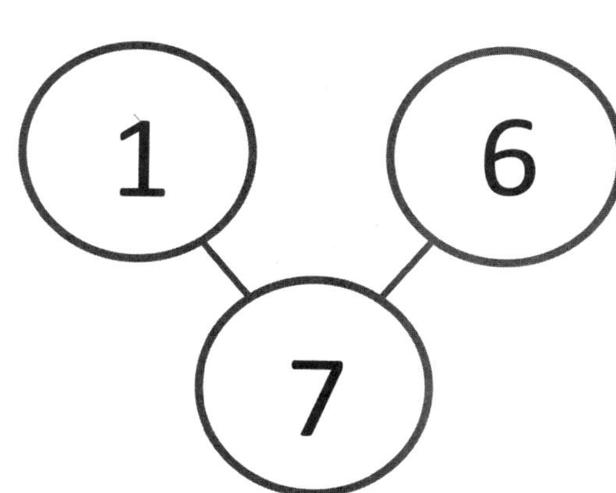
1
6
7

In each stick, color some cubes orange and the rest purple. Fill out the number bond to match. Tell a story about one of your number bonds to a friend.

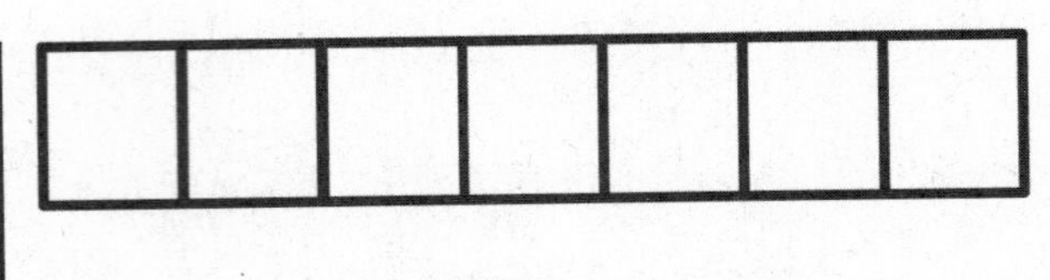

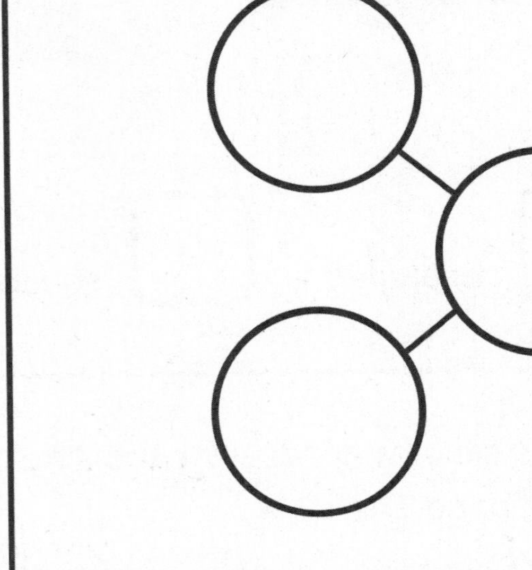

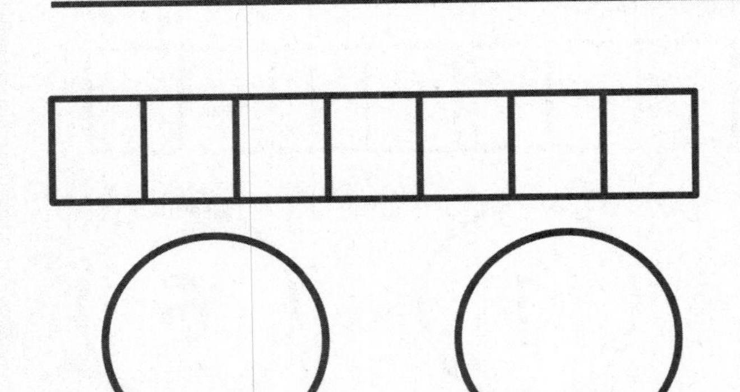

Draw a 7-stick, and use 2 colors to make 7. Make a number bond, and fill it in.

Take a 5-stick. Add 1 more cube.

How many cubes are in your stick now?

Draw a number bond to show how many cubes are in your stick now.

Draw

(Give students two linking cube 5-sticks, one each of two colors.) Add 1 more cube. How many are in your stick now? (7.) Add another cube. Now, how many cubes are in your stick? (8.) Take your 8-stick apart. Work with your partner to make two rows of cubes out of your stick. Make sure you have the same number of cubes in each row. How many cubes are in each row? (4.) Now, take your cubes, and make a tiny row of 2. Make another tiny row of 2 underneath. Keep going until all of your cubes are used up. How many cubes are in each row? (2.) How many tiny rows do you have? (4.) Talk to your partner about the ways you made your 8 look.

Name _______________ Date __________

Fill in the number bond to match the picture.

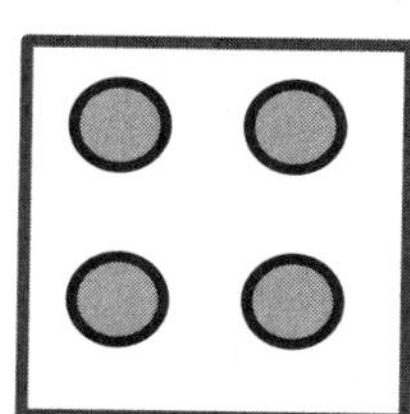
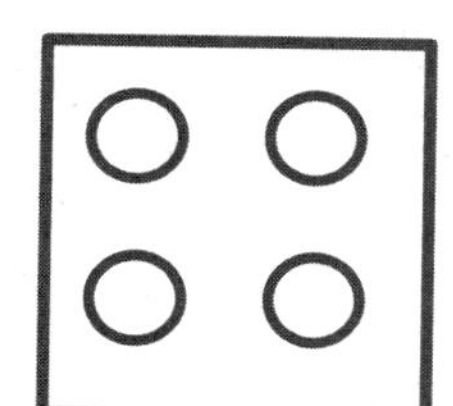
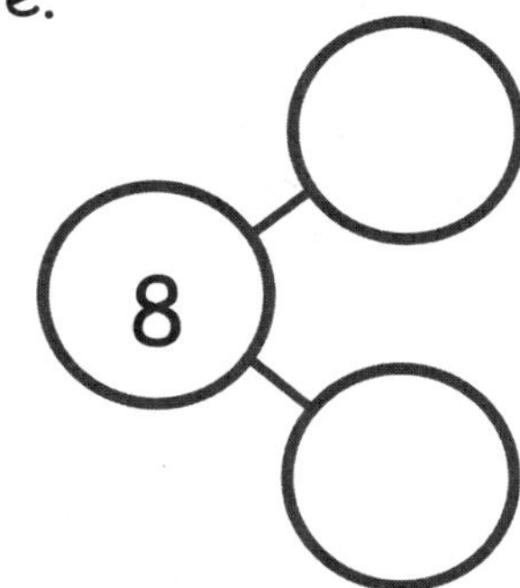

Draw some more dots to make 8 dots in all, and finish the number bond.

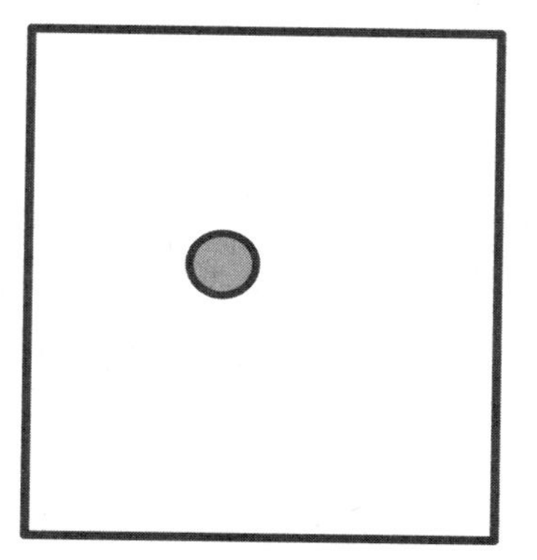
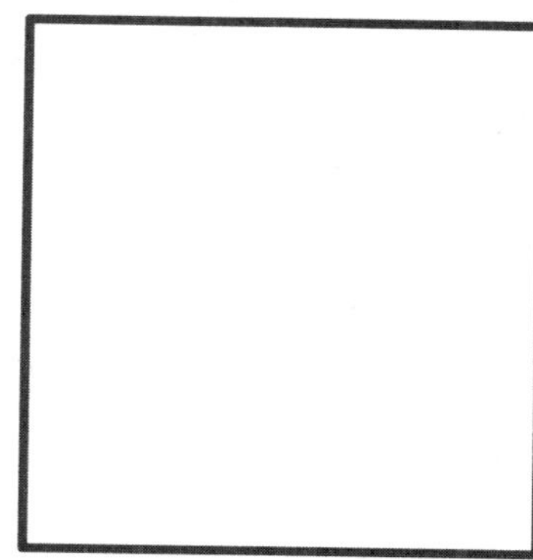
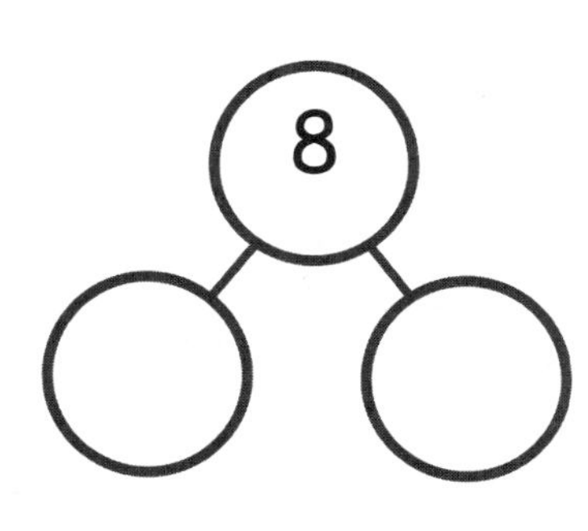

Draw 8 dots, some blue and the rest red. Fill in the number bond.

Blue Dots

Red Dots

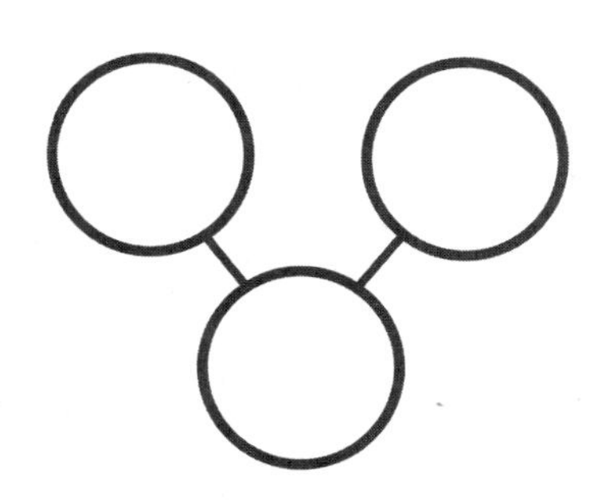

Draw a line to make 2 groups of dots. Fill in the number bond.

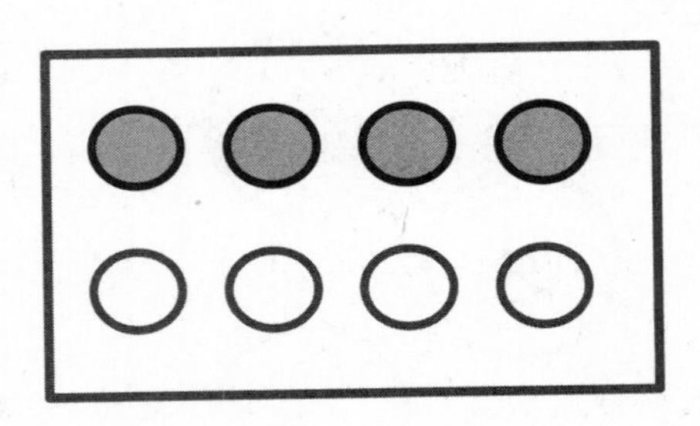

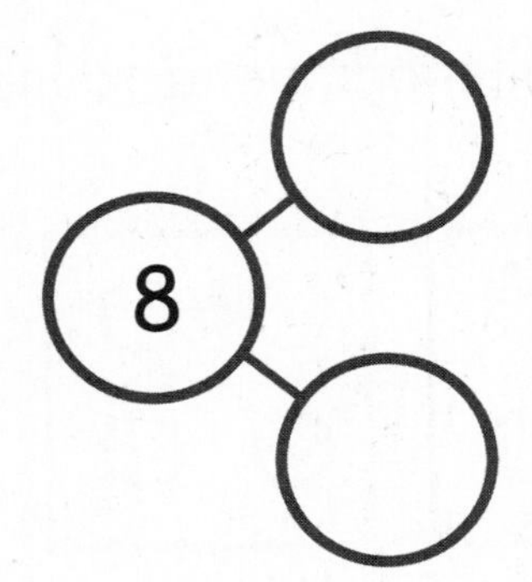

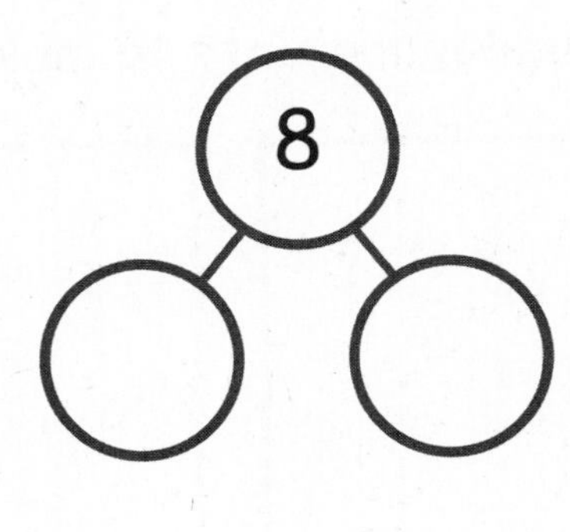

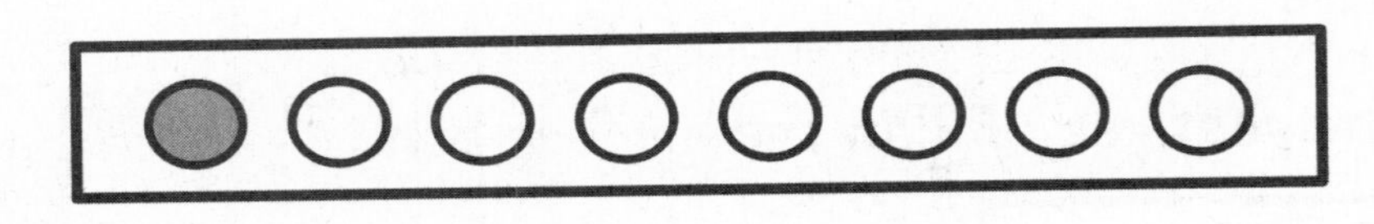

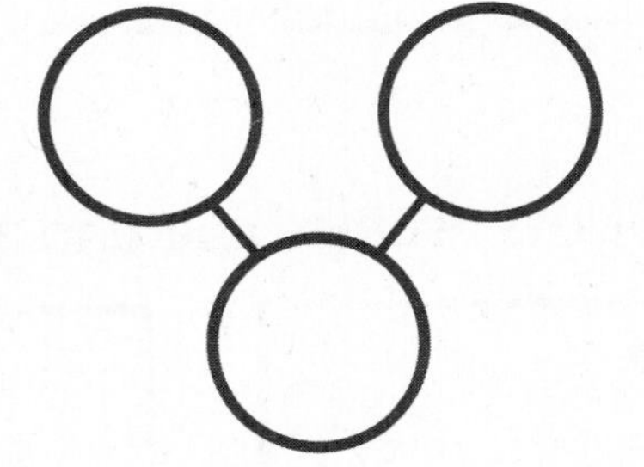

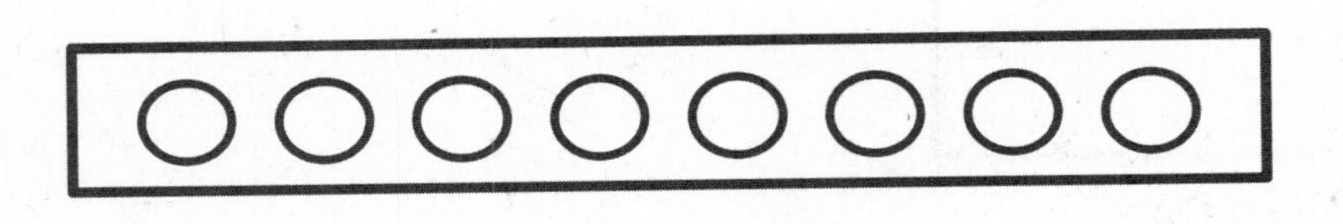

Play a game called Snap with a friend.

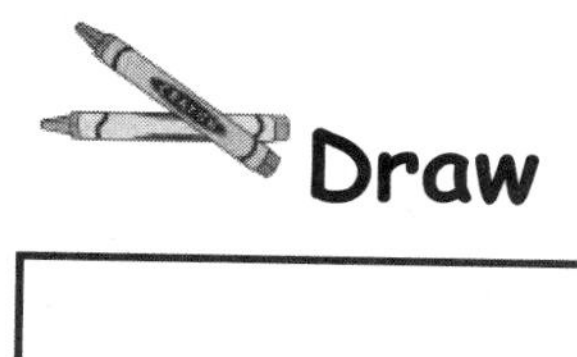

Draw

(Give students a 6-stick of linking cubes.) Take turns with a partner. Hold your 6-stick behind your back. When your partner says, "Snap!" break it into two parts. Show one of the parts while keeping the other behind your back. Can your partner guess the other part? Show the missing piece. Draw a number bond to show the two parts you made.

Name ______________________________ Date ______________

Fill in the number bond to match.

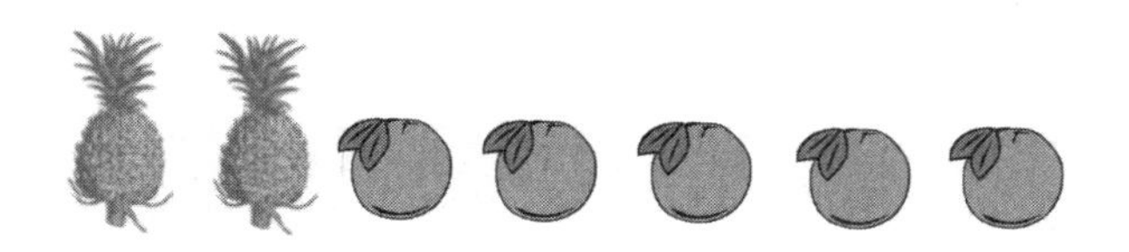

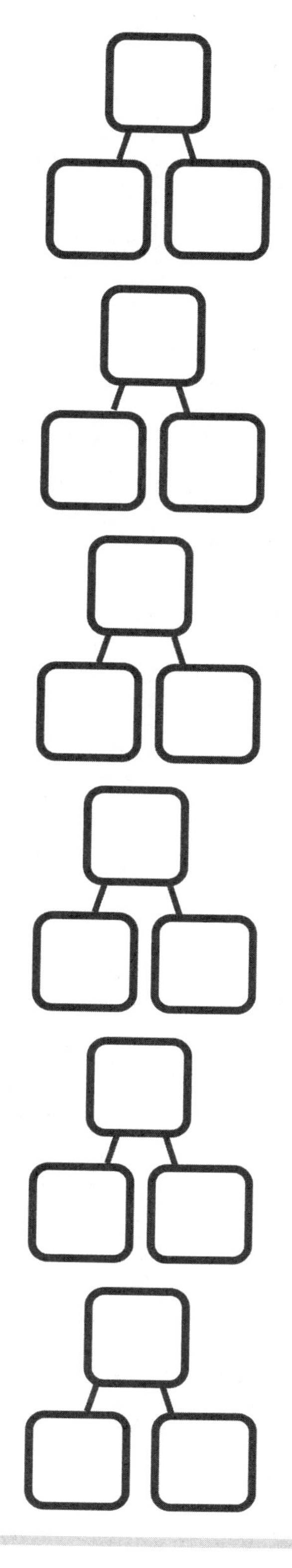

Color some of the faces orange and the rest blue. Fill in the number bond.

Color some of the faces orange and the rest blue. Fill in the number bond.

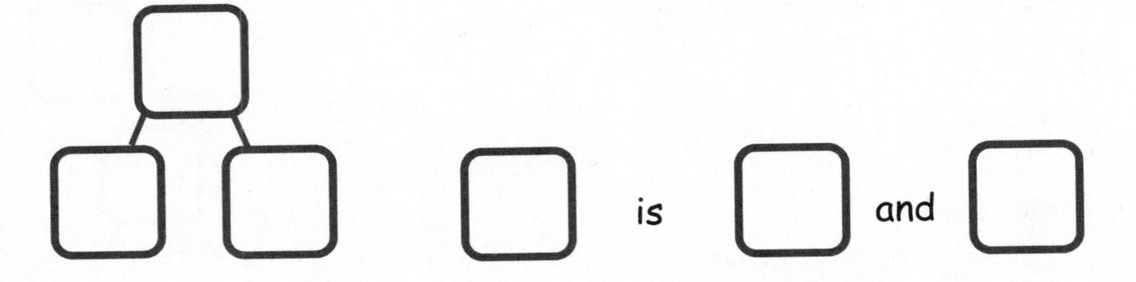

Color some of the faces orange and the rest blue. Fill in the number bond.

[] is [] and []

Lesson 10: Model decompositions of 6–8 using linking cube sticks to see patterns.

EUREKA MATH

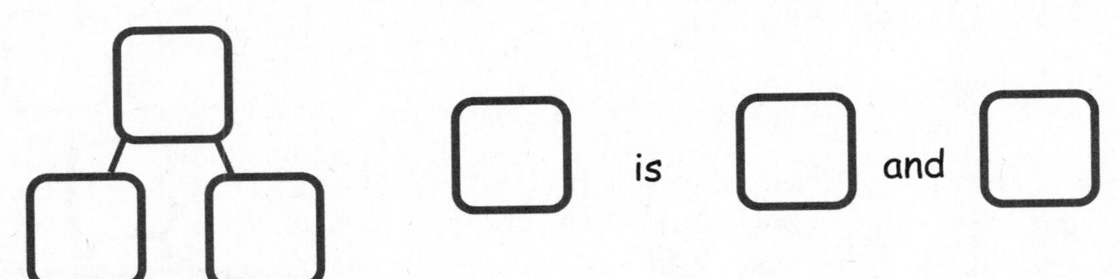

Nesim has 5 toy cars.

Awate has 3 toy cars.

How many cars do they have together?

Draw a picture to show Nesim's and Awate's cars. Draw a number bond to show Nesim's and Awate's cars. Tell a friend about your number bond.

Name ____________________ Date __________

These squares represent cubes. Draw a line to break the stick into 2 parts. Complete the number bond and number sentence.

6

6 is ☐ and ☐

7

☐ is ☐ and ☐

☐ is ☐ and ☐

☐ is ☐ and ☐

On the back of your paper, draw a cube stick with some red cubes and some blue cubes. Draw a number bond to match.

5 bees buzz around a flower.

2 more bees come to join them.

How many bees are there now?

Draw

Draw the bees and a number bond to go with the story. Talk to your partner about your picture.

Name ____________________ Date __________

5 boxes are colored. Color 3 more boxes to make 8. Complete the number bond.

8

5

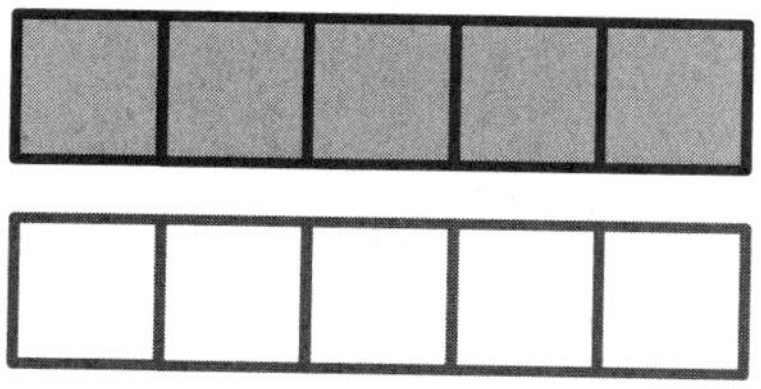

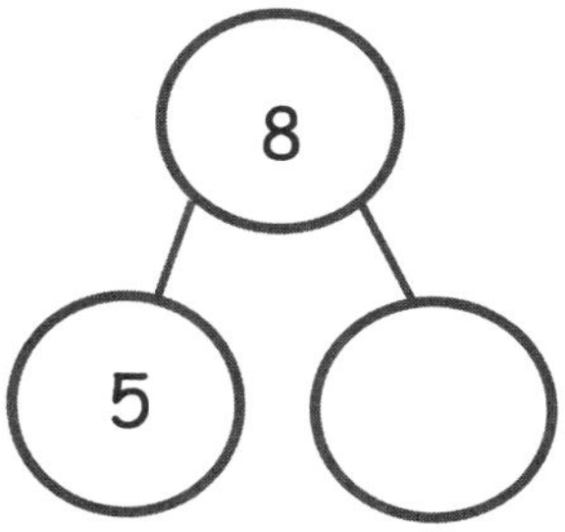

5 and ☐ more is 8

5 boxes are colored. Color more boxes to make 7. Complete the number bond.

5

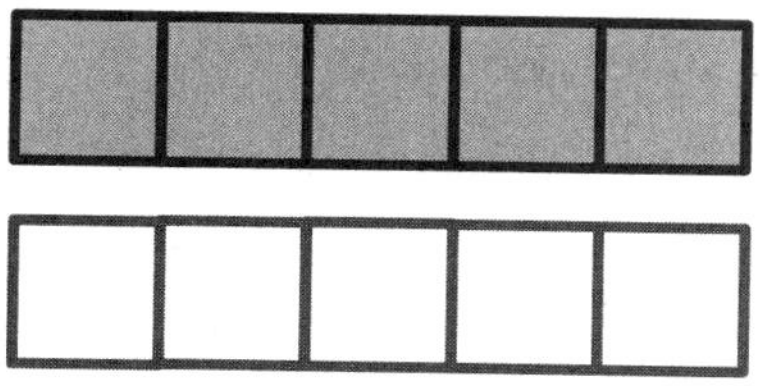

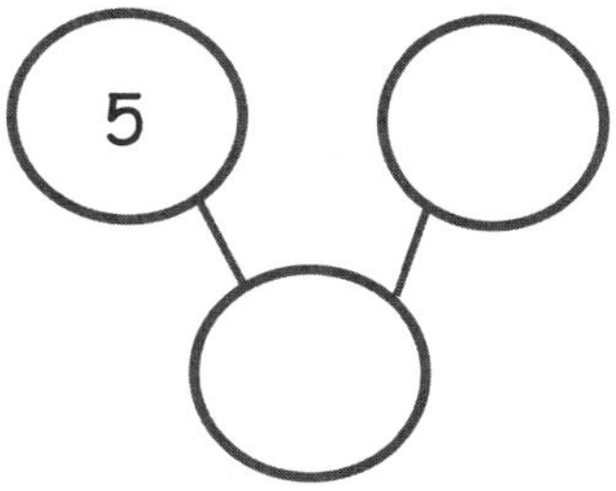

5 and ☐ more is ☐

Color 6 cubes. Complete the number bond.

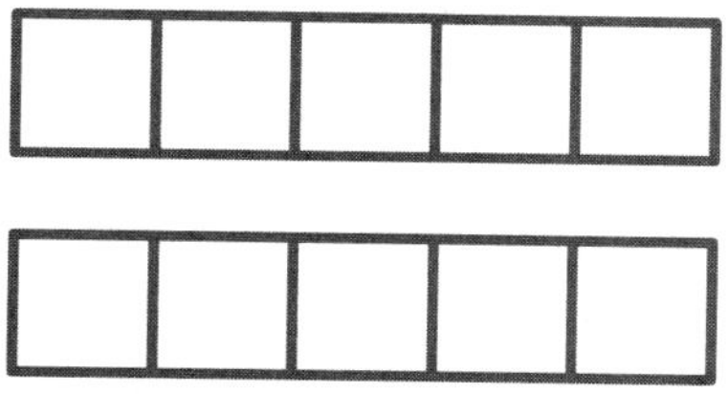

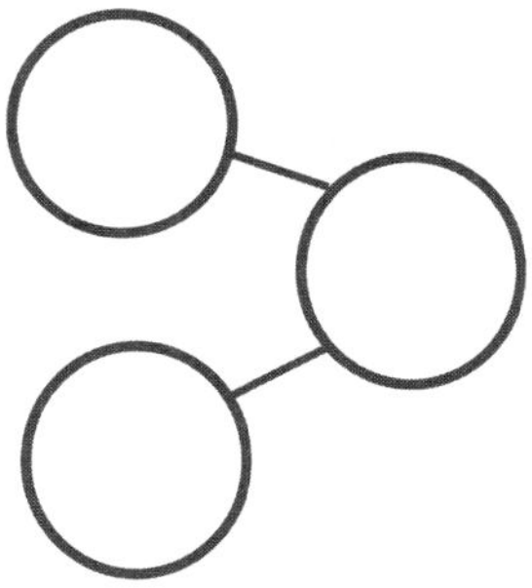

5 and ☐ more is ☐

Draw more to make 6. Complete the number bond.

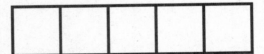

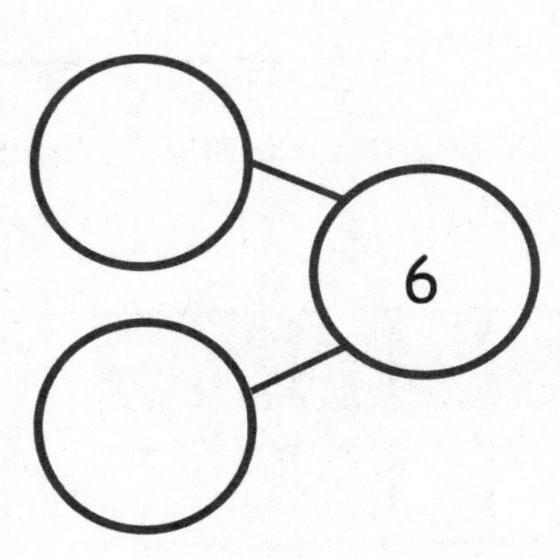

Draw more to make 7. Complete the number bond.

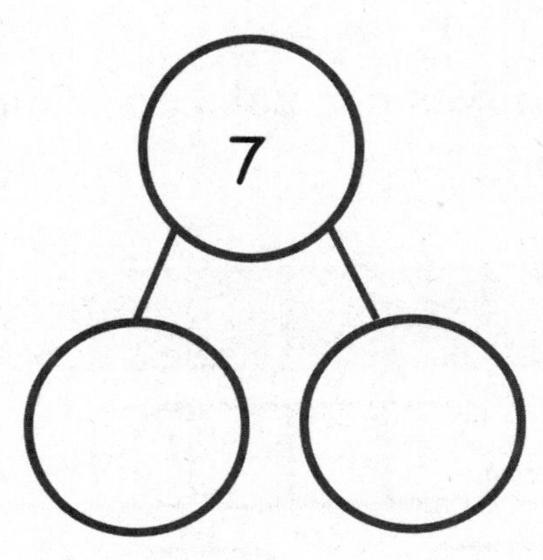

Draw more to make 8. Complete the number bond.

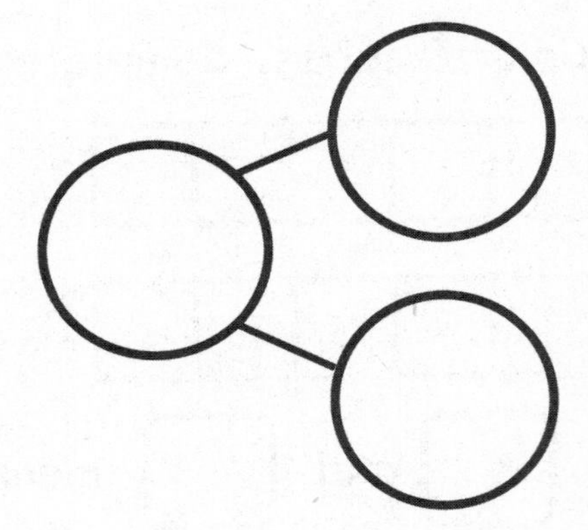

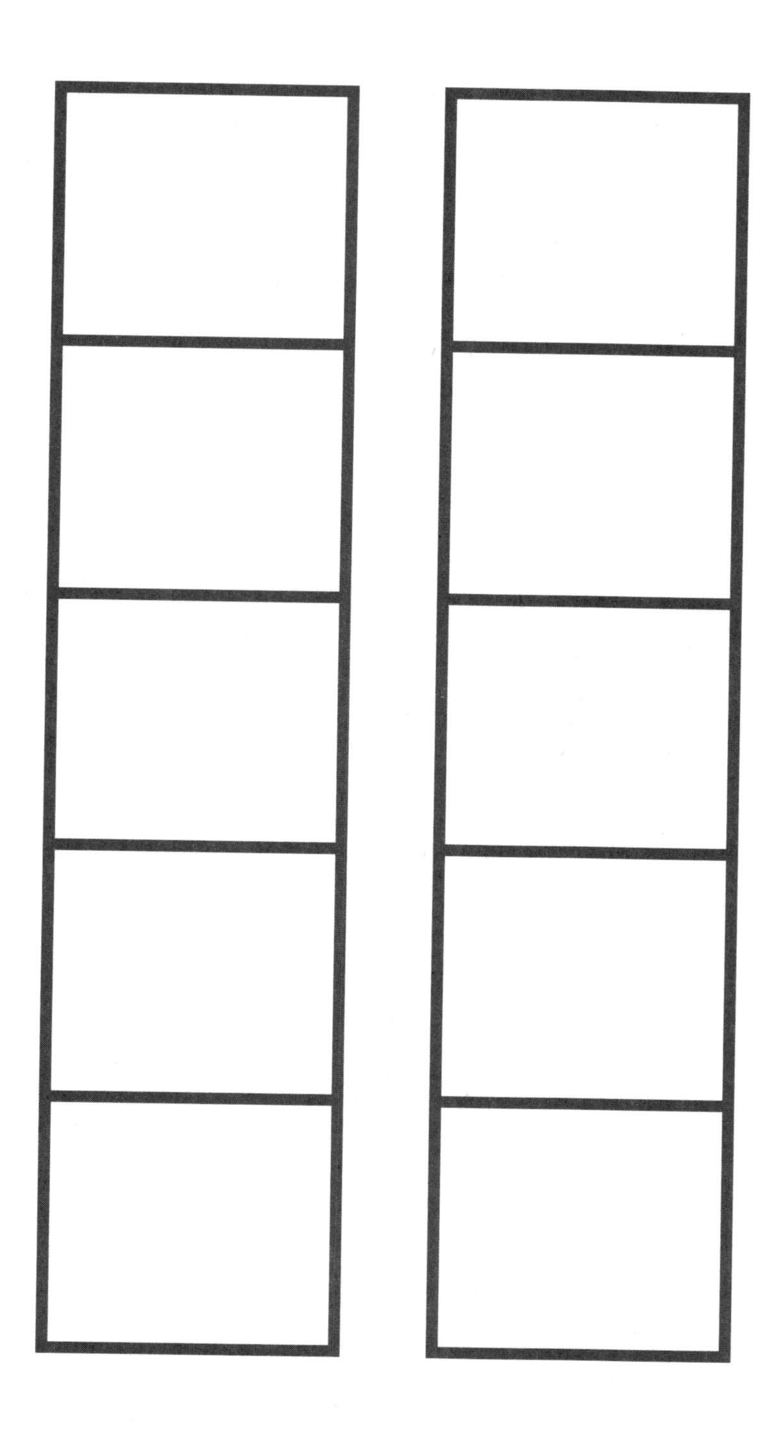

two 5-group mat

4 seals play in the water.

2 more seals come to play.

How many seals are playing now?

Draw

(Give students 6 linking cubes.) Use your linking cubes to show how many seals are playing. Draw a number bond to go with the story.

Name ______________________________ Date ______________

Fill in the number bond and number sentences.

There are 6 cornstalks. 5 cornstalks are in the first row. 1 cornstalk is in the second.

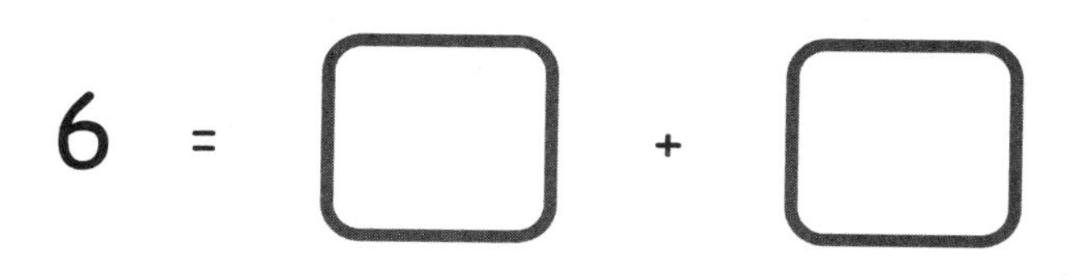

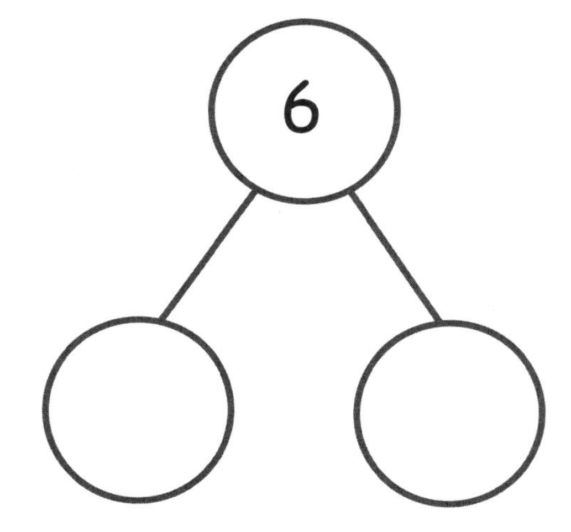

There are 6 cars on the road. 2 cars are big, and 4 are small.

☐ is ☐ and ☐

☐ = ☐ + ☐

3 geckos have black spots, and 3 geckos have no spots. There are 6 geckos.

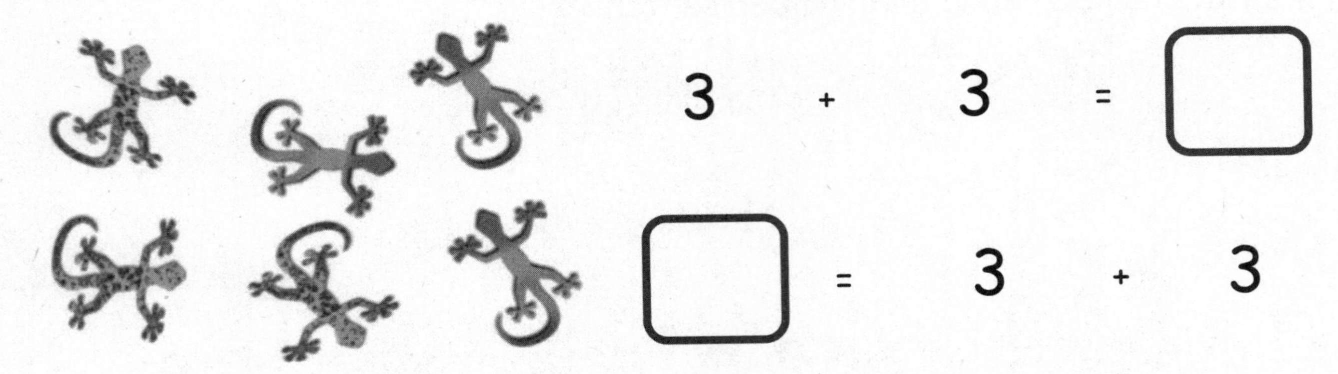

There are 6 monkeys. 4 monkeys are swinging on the tree, and 2 monkeys are taking a nap. Draw a picture to go with the story.

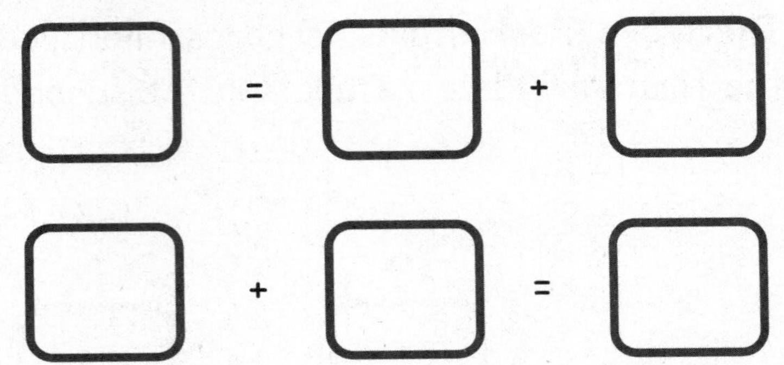

Create your own story, and tell your partner. Have your partner draw a picture of your story and create a number sentence to go with the picture.

Larry the train is pulling 7 cars.

3 cars are full.

4 cars are empty.

Draw

Draw the train, and make a number bond about your picture. Discuss your work with your partner.

Extension: Can you make a number sentence to go with your picture?

Name _______________________ Date ___________

There are 7 animals. There are 5 giraffes and 2 elephants.

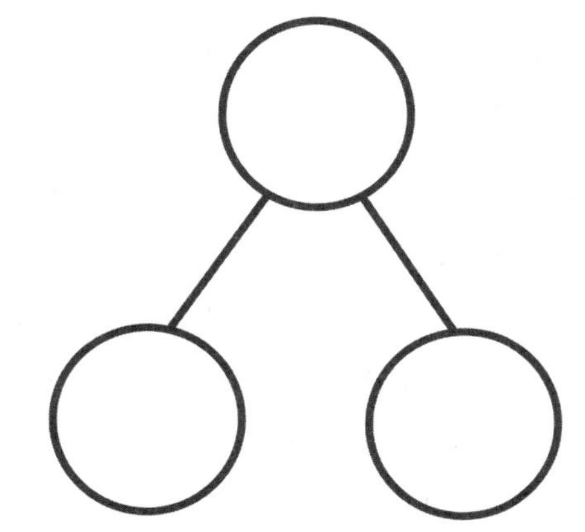

☐ = 5 + 2

At the store, there was 1 big bear and 6 small bears. There were 7 bears.

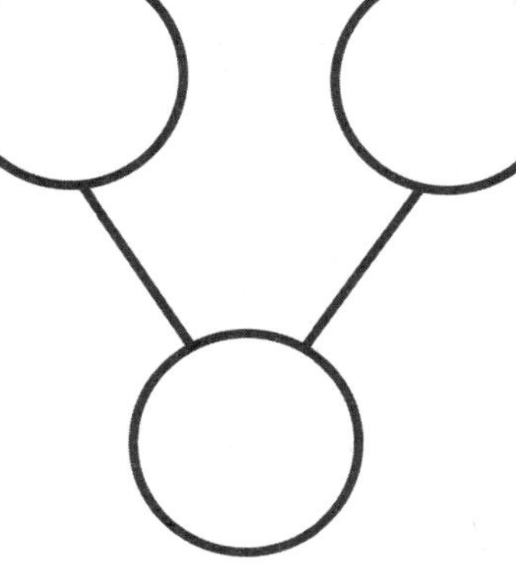

1 + 6 = ☐

☐ = 2 + 5

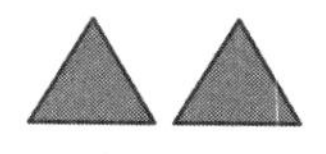
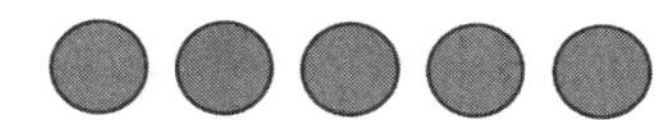
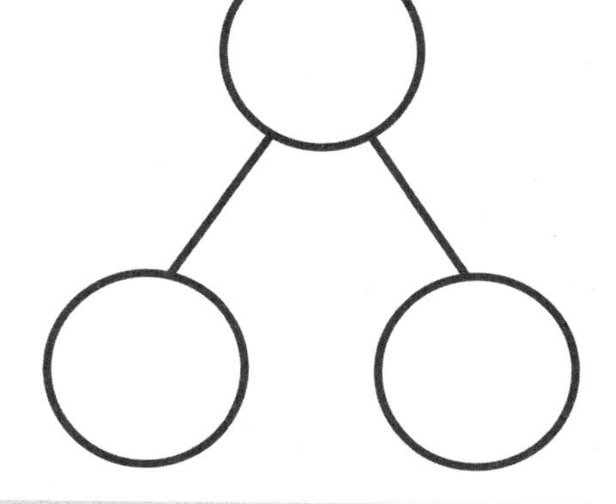

The squares below represent cubes.
4 gray cubes and 3 white cubes are 7 cubes.

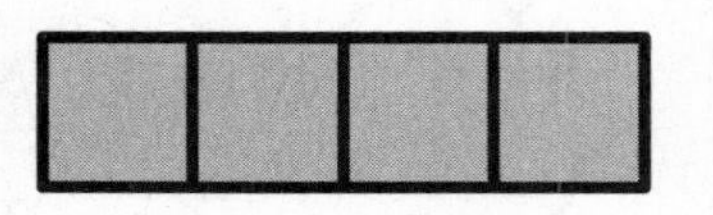
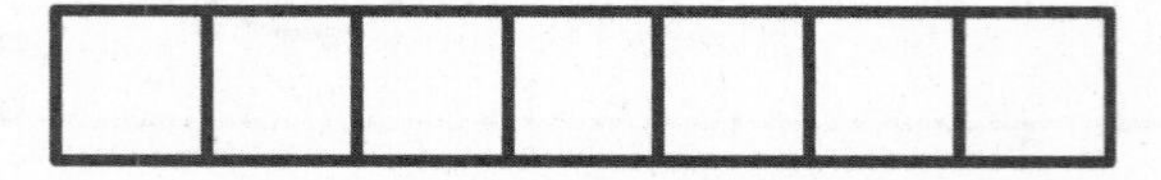
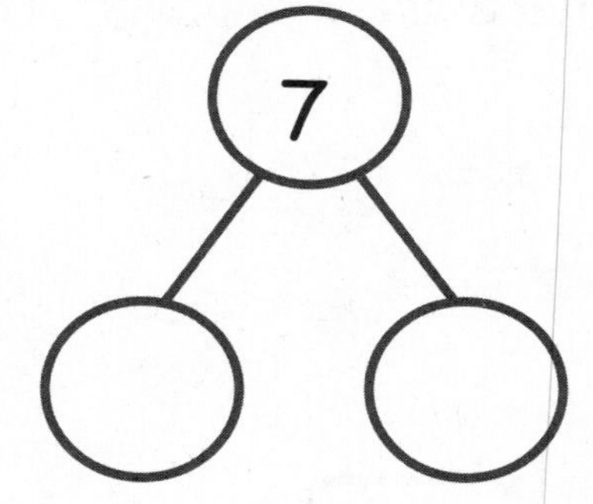

4 + 3 = ☐

Color the cubes to match the cubes above. Fill in the number sentence.

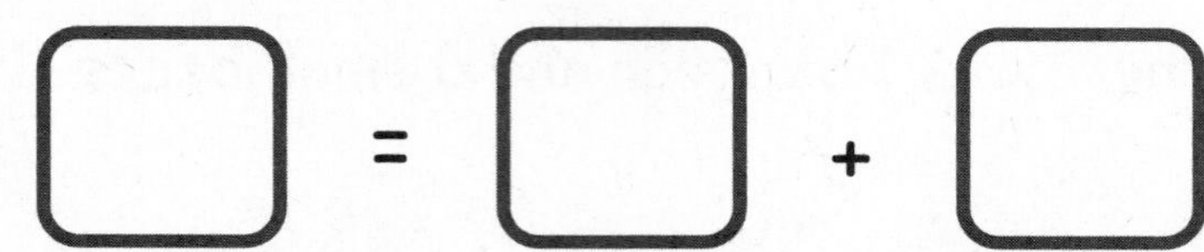

☐ = ☐ + ☐

Create your own story, and tell your partner. Have your partner draw a picture of your story and create a number sentence to go with the picture.

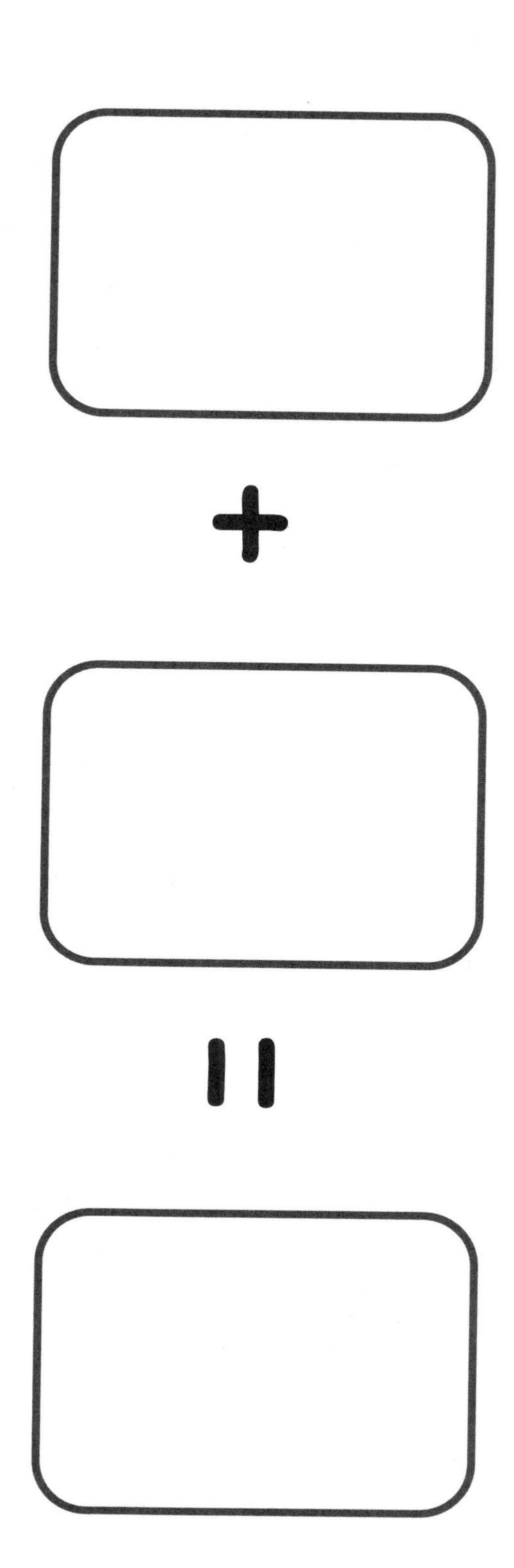

train

EUREKA MATH

Lesson 14: Represent decomposition and composition addition stories to 7 with drawings and equations with no unknown.

61

You are having a party!

You get 8 presents.

2 presents have stripes, and 6 presents have polka dots.

Draw

Draw the presents, and write the number sentences two different ways.

Write

Name ______________________ Date ____________

Fill in the number sentences.

There are 8 fish. There are 4 striped fish and 4 goldfish.

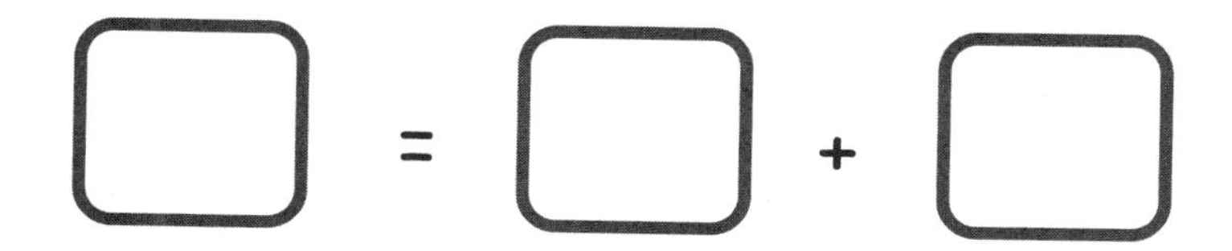

There are 8 shapes. There are 5 triangles and 3 diamonds.

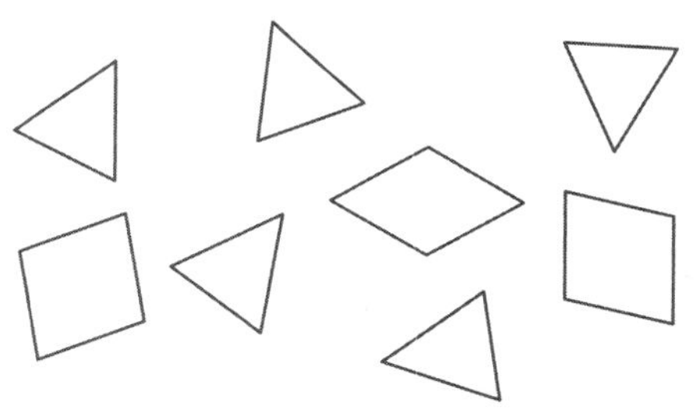

There are 6 stars and 2 moons.
There are 8 shapes.

There are 8 shapes. Count and circle the squares. Count and circle the triangle.

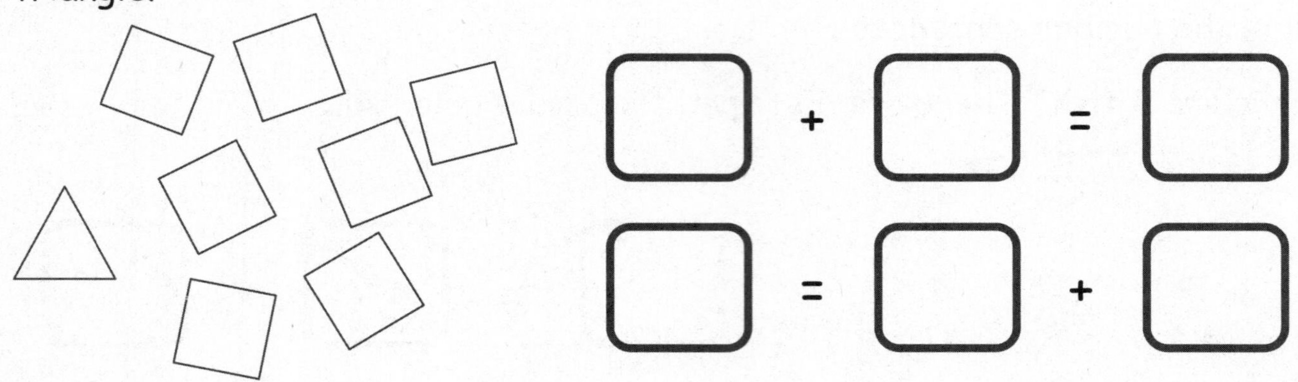

There are 8 flowers. Some flowers are yellow, and some flowers are red. Draw a picture to go with the story.

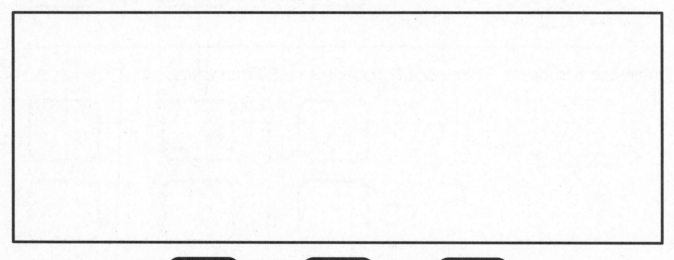

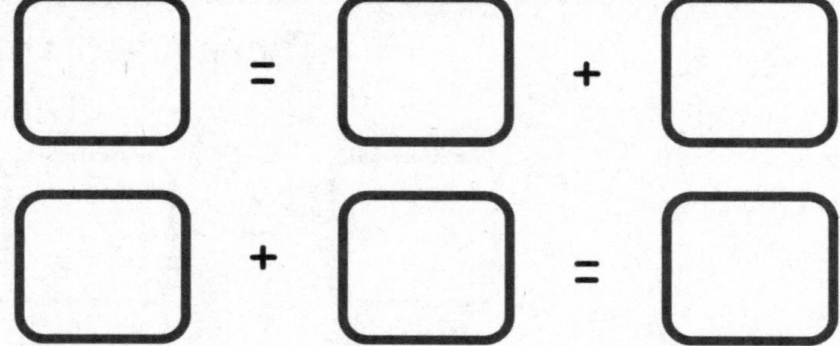

Create your own story, and tell your partner. Have your partner draw a picture of your story and create a number sentence to go with the picture.

3 airplanes were flying in the air.

3 more airplanes came to join them.

How many airplanes were flying in the air?

Draw

(Give students 10 linking cubes.) Use your linking cubes to show how many airplanes were flying in the air. Write a number sentence to go with the story. Share your number sentence with a partner.

Write

Name ______________________________ Date ______________

There are 4 snakes sitting on the rocks. 2 more snakes slither over. How many snakes are on the rocks now? Put a box around all the snakes, trace the mystery box, and write the answer inside it.

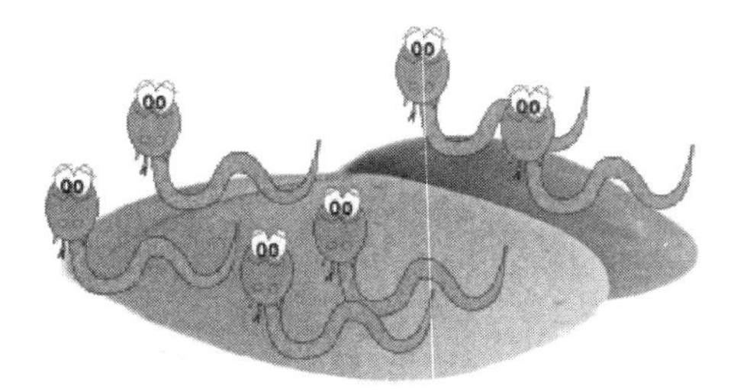

$4 + 2 = \square$

There are 5 turtles swimming. Draw 2 more turtles that come to swim. How many turtles are swimming now? Draw a box around all the turtles, draw a mystery box, and write the answer.

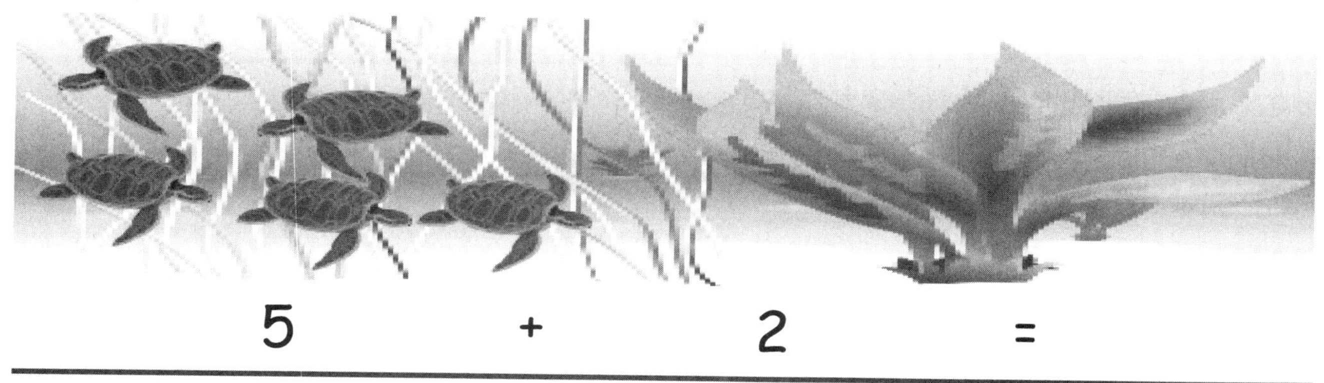

$5 + 2 =$

Today is your birthday! You have 7 presents. A friend brings another present. Draw the present. How many presents are there now? Draw a mystery box, and write the answer inside it.

$7 + 1 =$

Listen and draw. There were 6 girls playing soccer. A boy came to play. How many children were playing soccer then? Draw a box around all the children.

6 + 1 = ☐

Listen and draw. There were 3 frogs on a log. 5 more frogs hopped onto the log. How many frogs were on the log then? Draw a box around the frogs, and box the answer.

3 + 5 =

Marissa is playing with shapes.

She has 5 triangles and 2 circles.

Draw the shapes, and write a number sentence.

Draw

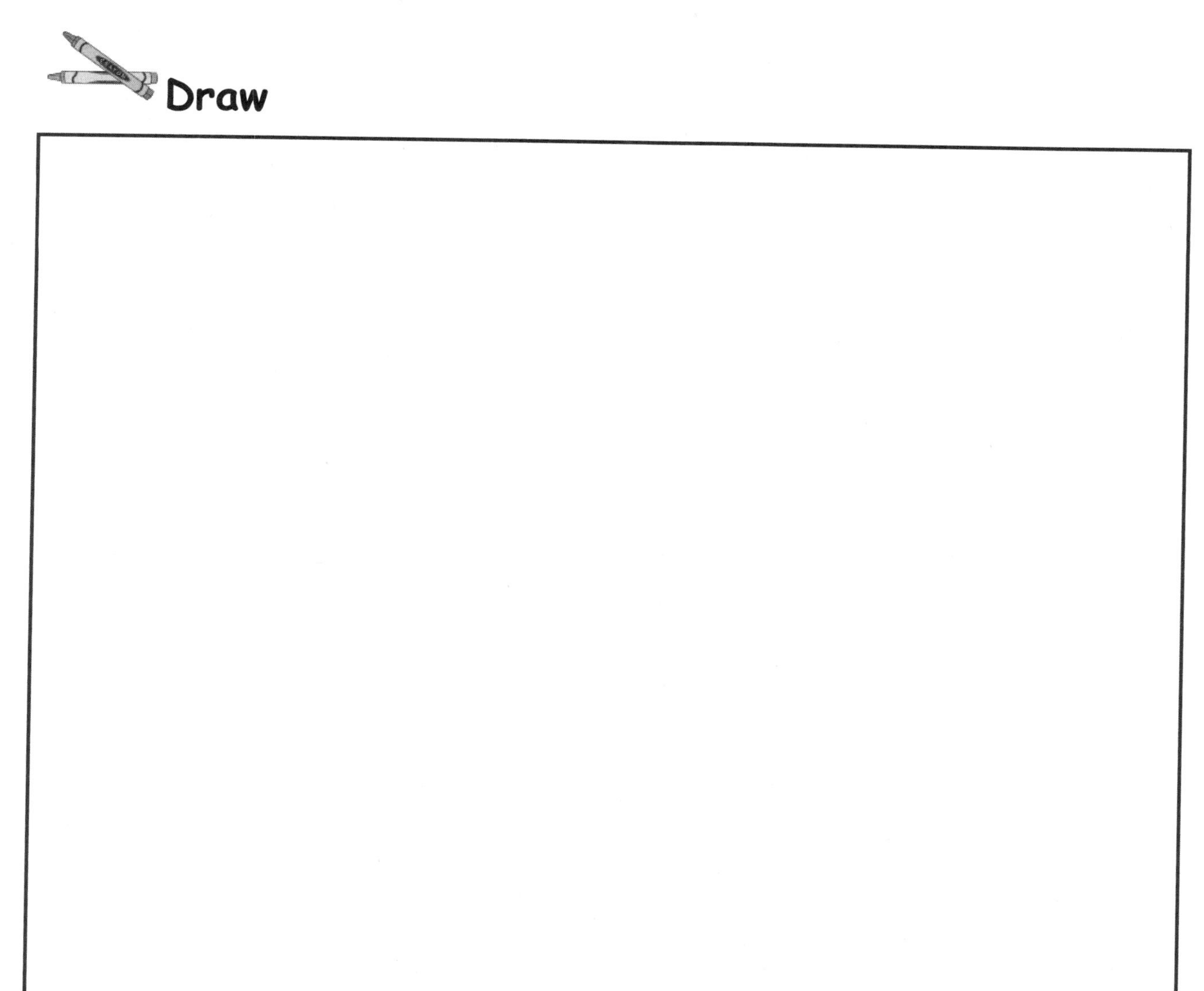

Talk to your partner about your picture and number sentence.

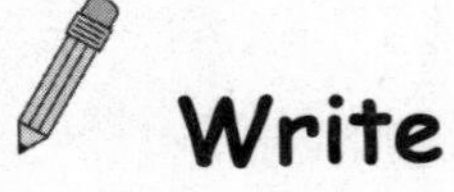

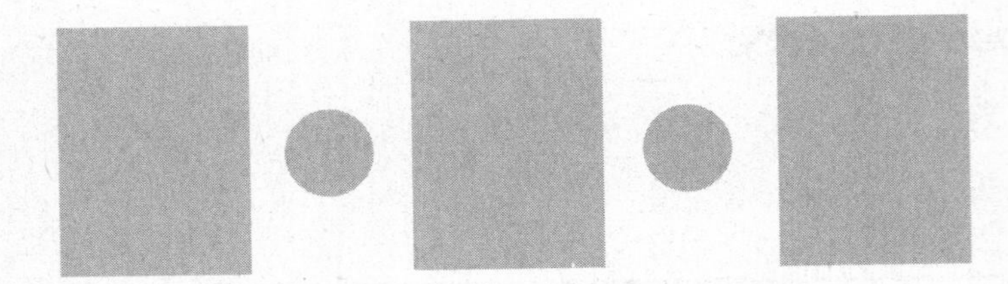

Name ______________________________ Date ______________

There are 4 green balloons and 3 orange balloons in the air. How many balloons are in the air? Color the balloons to match the story, and fill in the number sentences.

Dominic has 6 yellow star stickers and 2 blue star stickers. How many stickers does Dominic have? Color the stars to match the story, and fill in the number sentences.

There are 5 big robots and 1 little robot. How many robots are there? Fill in the number sentences.

Listen and draw. Charlotte is playing with pattern blocks. She has 3 squares and 3 triangles. How many shapes does Charlotte have?

☐ + ☐ = ☐

☐ = ☐ + ☐

Listen and draw. Gavin is making a tower with linking cubes. He has 5 purple and 3 orange cubes. How many linking cubes does Gavin have?

☐ + ☐ = ☐

☐ = ☐ + ☐

tree and sun

Sam bought 8 pieces of fruit.

Some were apples and some were oranges.

Draw a plate, and show Sam's fruit on the plate.

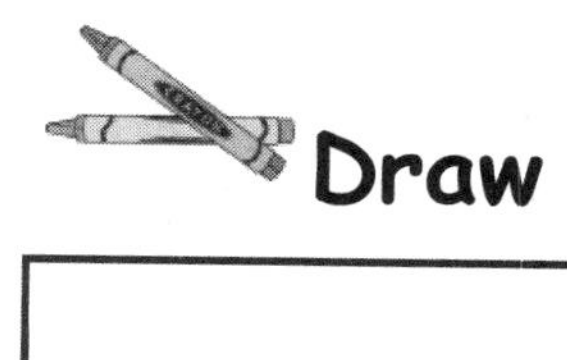

Draw

Show your picture to a friend. Do your plates look the same? Make a number bond and number sentence to go with your picture.

Write

Name ________________________________ Date ______________

Devin has 6 Spiderman pencils. He put some in his desk and the rest in his pencil box. Write a number sentence to show how many pencils Devin might have in his desk and pencil box.

6 = ☐ + ☐

Shania made 7 necklaces. She wore some of the necklaces and put the rest in her jewelry box. Use the linking cubes to help you think about how many necklaces Shania might have on and how many are in her jewelry box. Then, complete the number sentences.

☐ + ☐ = ☐

☐ = ☐ + ☐

Tommy planted 8 flowers. He planted some in his garden and some in flowerpots. Draw how Tommy may have planted the flowers. Fill in the number sentences to match your picture.

Create your own story, and draw a picture. Fill in the number sentences. Tell your story to a friend.

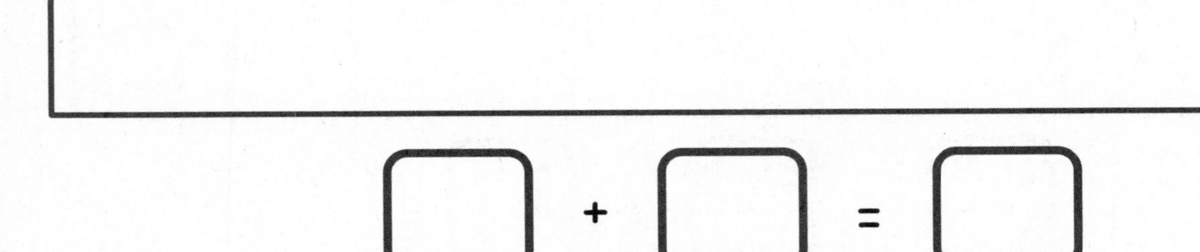

Make 5 little pieces of cheese out of your clay.

2 mice each stole a piece of cheese.

Draw

(Give students a ball of clay.) Take away pieces to show that the mice ate them. How many pieces are left? Act out the story again with 4 pieces of cheese. How many are left? Talk about the mice and cheese with your partner. Did your partner have the same number of pieces left each time? What do you think would happen if you only had 3 pieces of cheese?

Name ______________________________ Date ______________

The cat ate 3 mice. Cross out 3 mice. Write how many mice are left.

The fish ate 2 worms. Cross out 2 worms. Write how many worms are left.

The frog ate 5 flies. Cross out 5 flies. Write how many flies are left.

The monkey ate 4 bananas. Cross out 4 bananas. Write how many bananas are left.

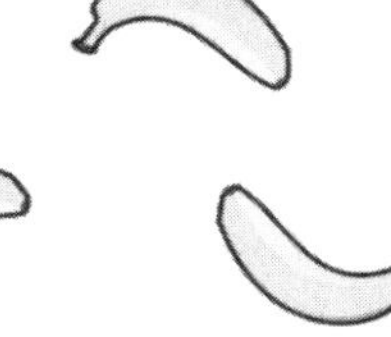

Draw 6 balls. The boy kicked 3 balls down the hill. How many balls does he have left?

There are 5 butterflies flying around the flower. Draw them. 1 of the butterflies flew away, so cross it out. How many butterflies are left?

Draw 5 monkeys jumping on the bed.

Decide how many monkeys stayed on the bed.

Cross off the monkeys who fell off and bumped their heads.

Draw

Share your picture with your partner. How many monkeys did you start with? How many did you take away? How many were left? How is your number story different from your partner's?

Name ______________________ Date ____________

Draw a line from the picture to the number sentence it matches.

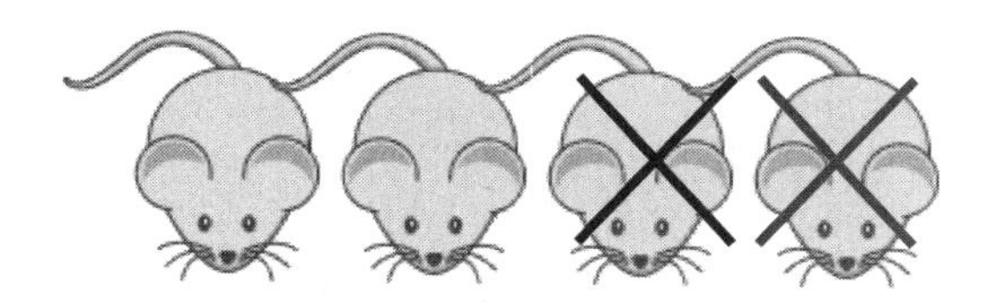

3 - 1 = 2

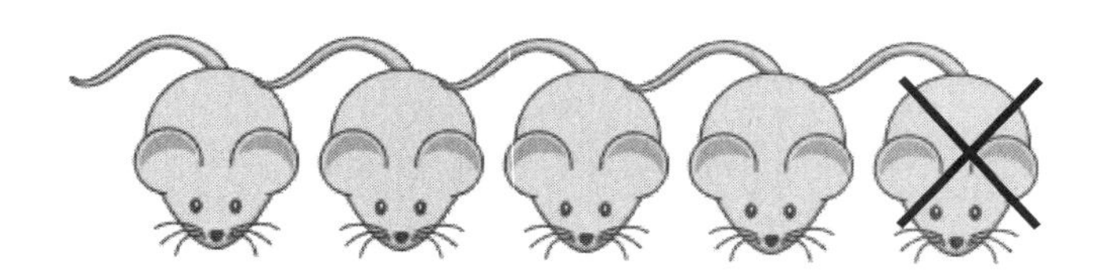

5 - 4 = 1

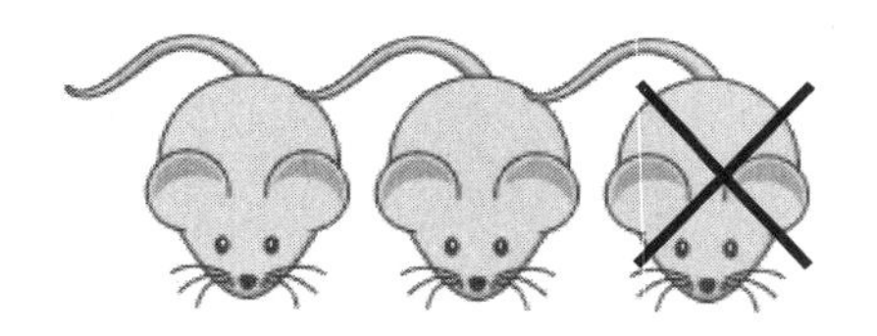

4 - 2 = 2

5 - 1 = 4

Pick 1 mouse picture, and tell a story to your partner. See if your partner can pick the picture you told the story about.

Cross out the bears to match the number sentences.

6 - 1 = 5

7 - 2 = 5

6 - 4 = 2

7 - 3 = 4

8 - 1 = 7

8 - 2 = 6

5 frogs were sitting by the pond.

2 frogs hopped into the pond.

How many frogs were still sitting by the pond?

Draw

Draw the frogs. Cross out the frogs in your picture to show the ones who hopped into the pond. Talk to your partner about the story. How can you write about your story in a number sentence?

Name ______________________________ Date ______________

Tyler bought a cone with 4 scoops. He ate 1 scoop. Cross out 1 scoop. How many scoops were left?

4 - 1 = ☐

Eva ate ice cream, too. She ate 2 scoops. How many scoops were left?

4 - 2 = ☐

There were 4 bottles. 3 of them broke. How many bottles were left?

4 - 3 = ☐

Anthony had 5 erasers in his pencil box. He dropped his pencil box, and 4 erasers fell on the floor. How many erasers are in Anthony's pencil box now? Draw the erasers, and fill in the number sentence.

5 - 4 = ☐

Tanisha had 5 grapes. She gave 3 grapes to a friend. How many grapes does Tanisha have now? Draw the grapes, and fill in the number sentence.

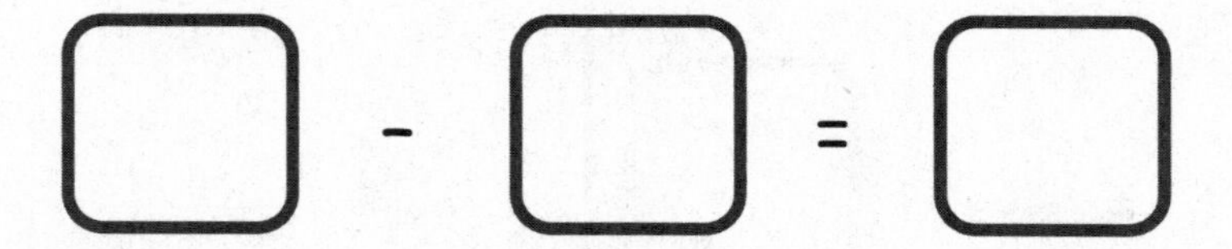

Play Snap with a friend.

Draw

(Give students a 6-stick of linking cubes.) Take turns with a partner. Hold your 6-stick behind your back. When your partner says, "Snap!" break it into two parts. Show one of the parts while keeping the other behind your back. Can your partner guess the other part? Show the missing piece. Draw a number bond to show the two parts you made. Can you and your partner think of a take away number sentence to tell about the snap?

Name ____________________ Date ____________

Fill in the number bonds.

Cross out 1 hat.

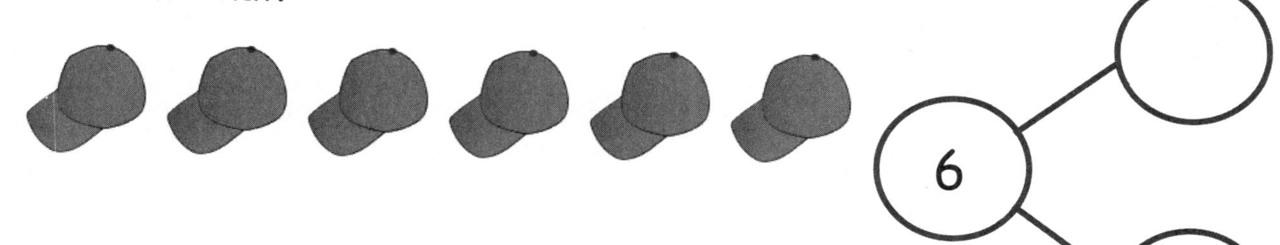

6 - 1 = 5

Cross out 5 snowflakes.

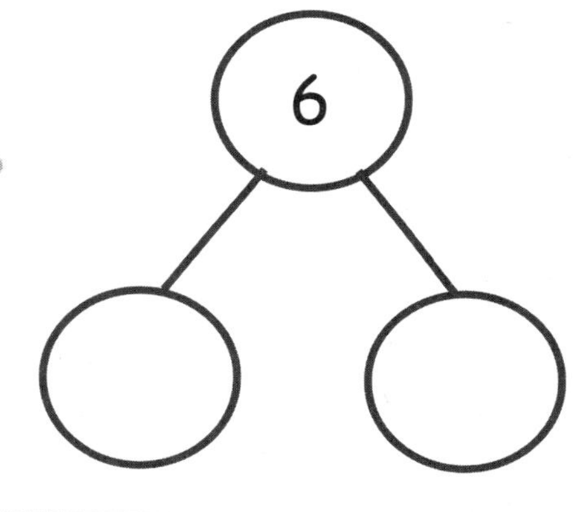

6 - 5 = 1

Cross out 2 snowflakes.

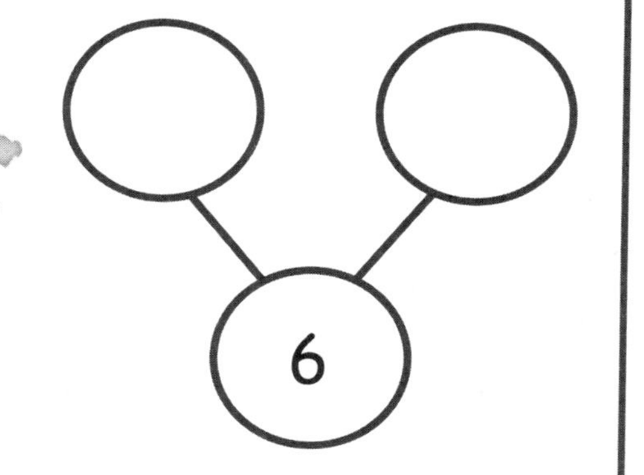

6 - 2 = 4

Fill in the number sentences and the number bonds.

Take away 3 hats.

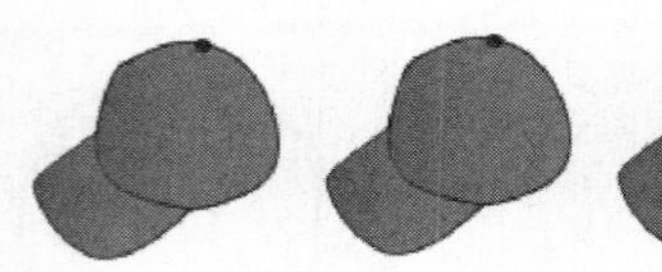
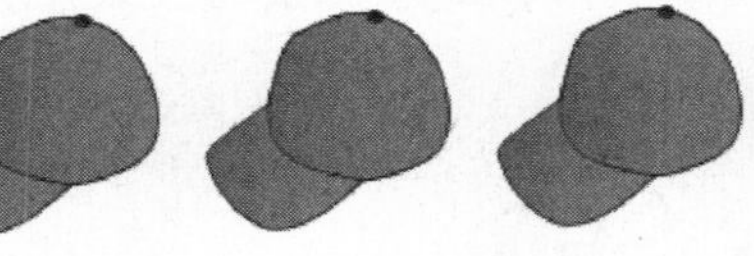
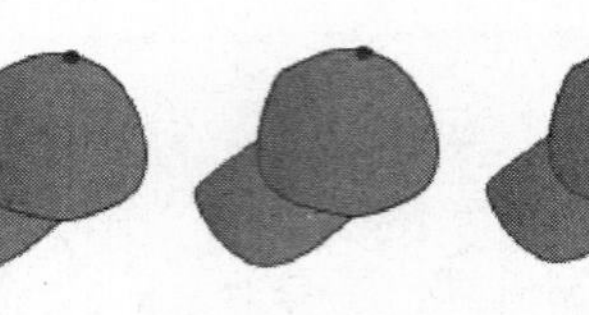

6 - ☐ = ☐

Take away 4 cubes.

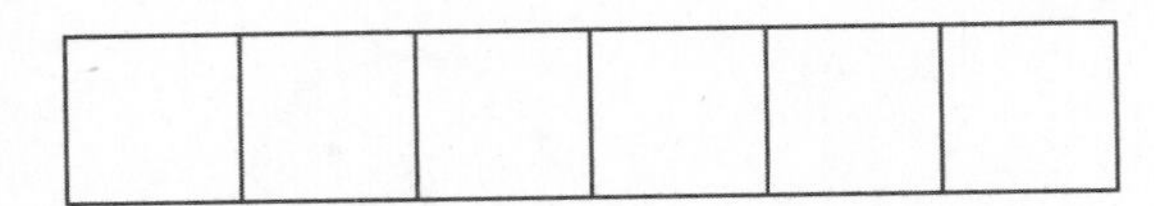
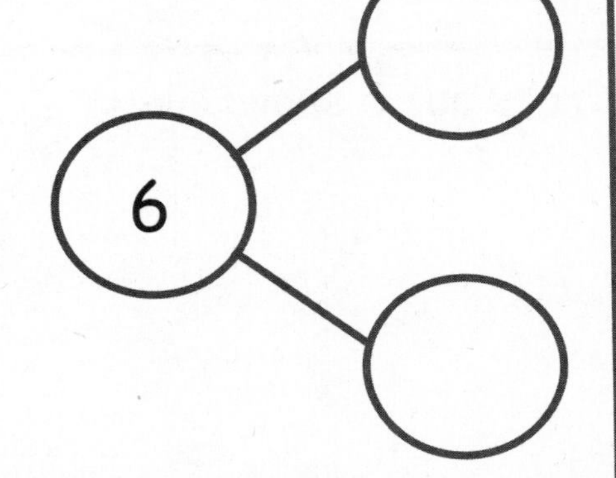
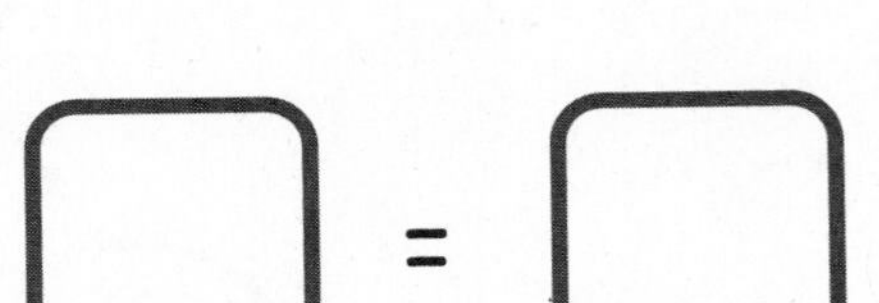

6 - ☐ = ☐

Draw 6 circles in a 5-group. Take away 2 circles.

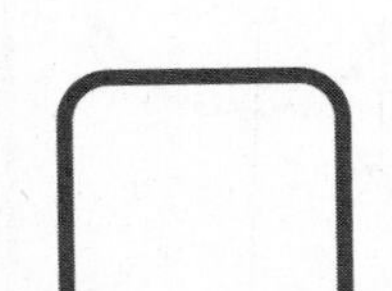

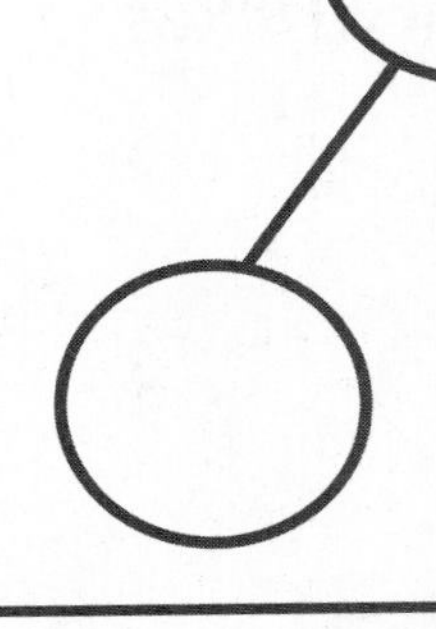

☐ - ☐ = ☐

Noah had 7 red balloons.

2 balloons popped.

Draw Noah's balloons.

Draw

How would you show that 2 balloons popped in your picture? Make a number sentence to go with your picture. Draw a number bond to go with your picture.

Write

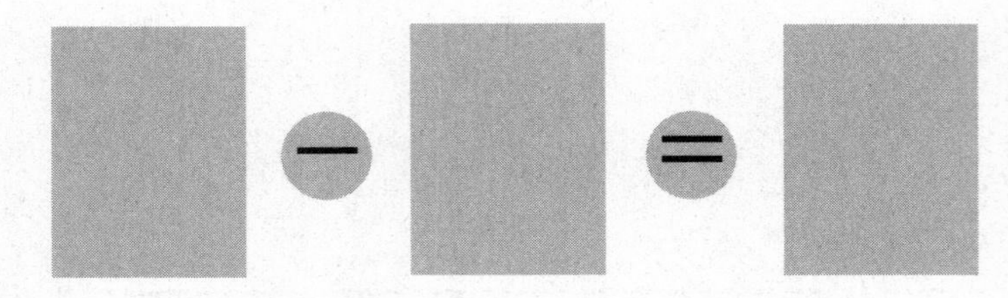

Name ______________________________ Date ______________

Say the number sentence. Fill in the blanks. Cross out the number.
Cross out 2 dots.

$7 - 2 = \square$

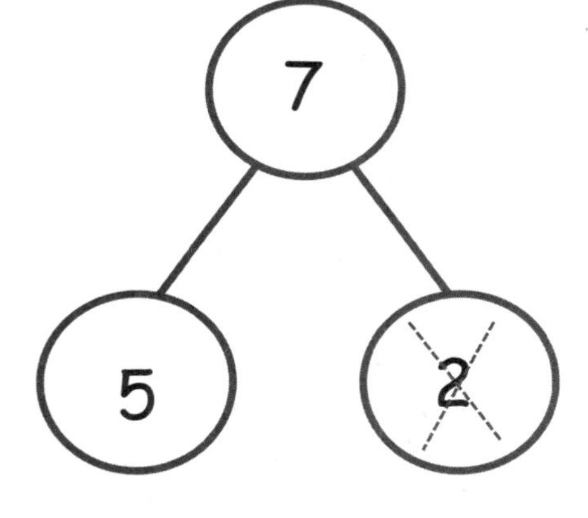

Cross out 5 dots.

$7 - 5 = \square$

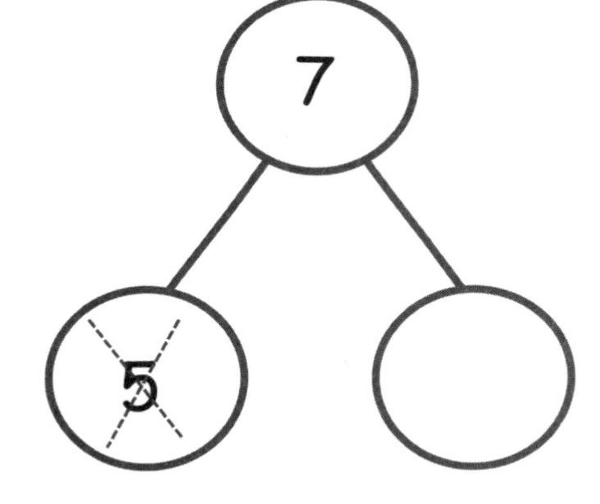

Cross out 4 dots.

$7 - 4 = \square$

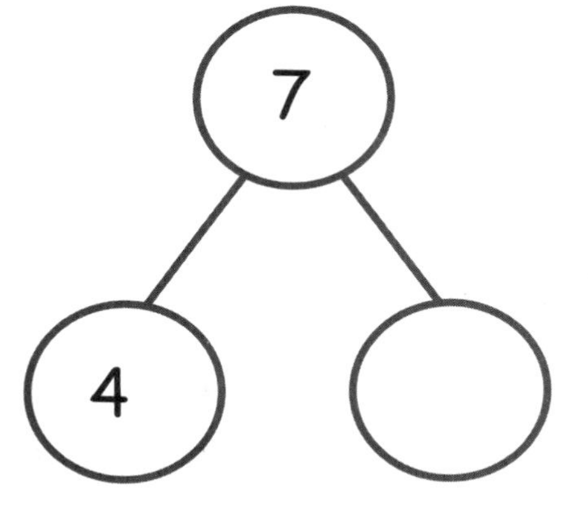

Draw and fill in the number bond and number sentence.

Draw 7 dots. Cross out 2 dots.

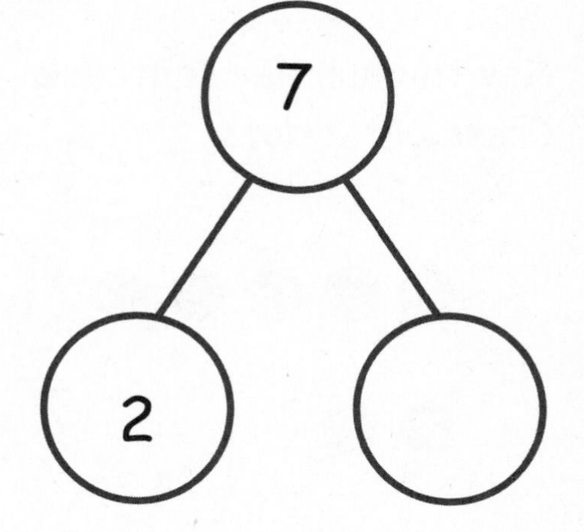

Draw 7 dots in a 5-group. Cross out 3 dots.

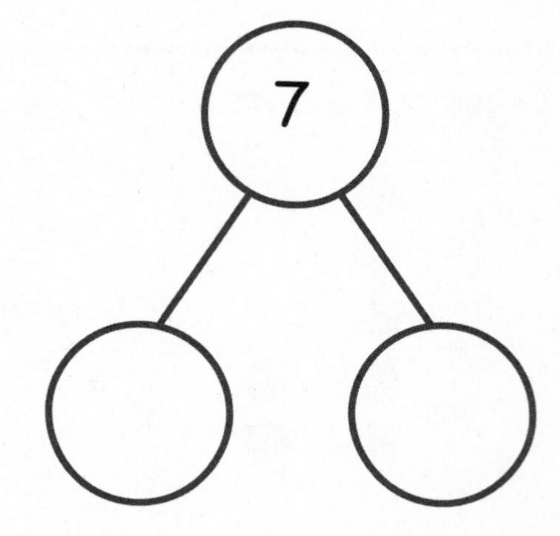

7 - ☐ = ☐

Draw 7 dots in a 5-group. Cross out 4 dots.

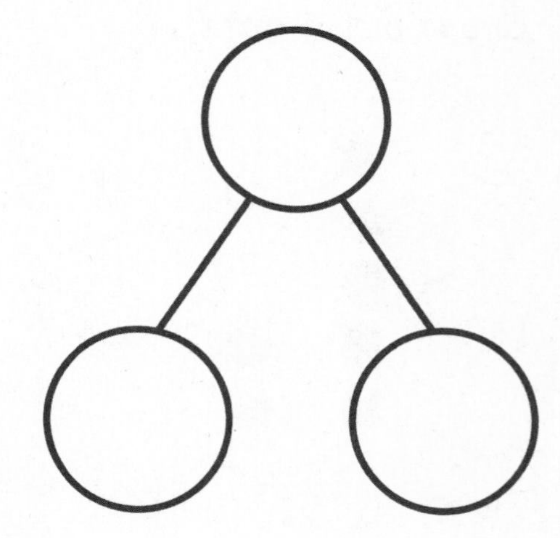

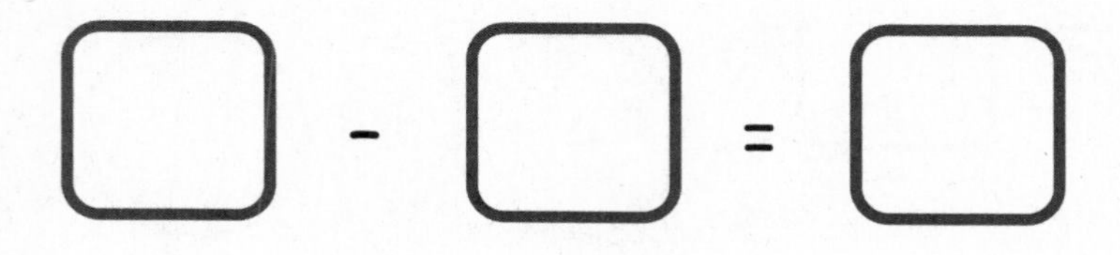

Robin had 8 cats in her house.

3 of the cats went outside to play.

How many cats were still in the house?

Draw

Draw Robin's cats. Use your picture to help you draw a number bond about the cats. Can you make a number sentence to tell how many cats were still inside the house?

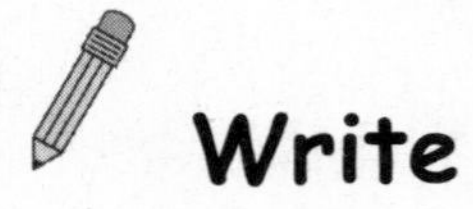

Name ______________________________ Date ______________

Fill in the number sentences and number bonds.

Put an X on 3 dots.

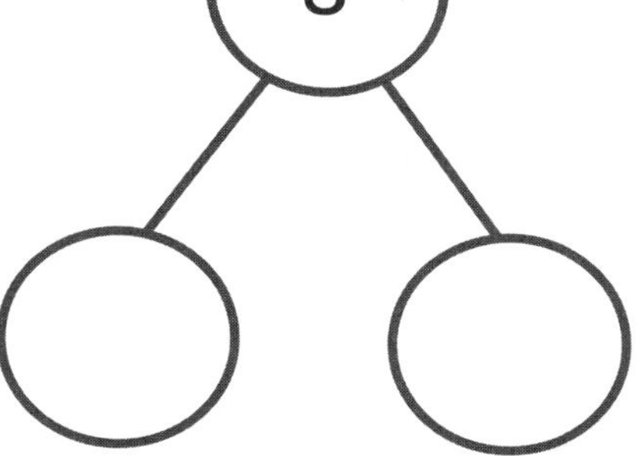

8 – ☐ = ☐

Put an X on 5 dots.

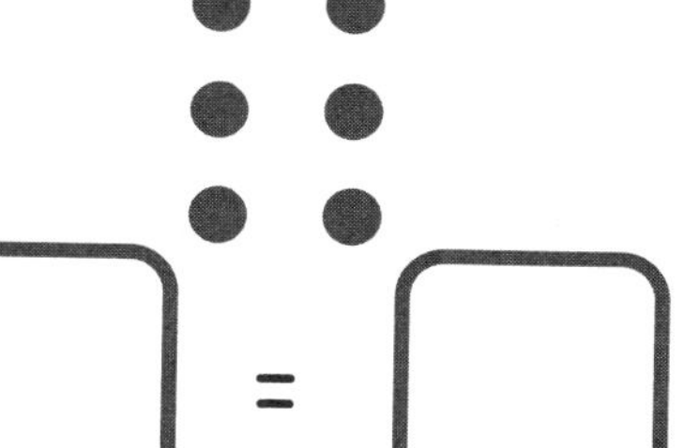

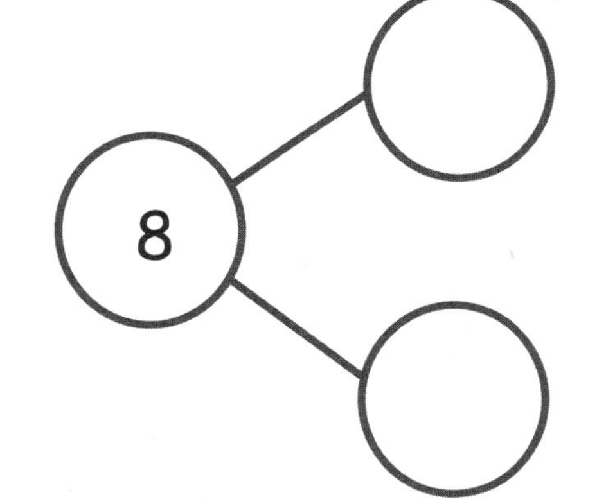

☐ – ☐ = ☐

Put an X on some dots.

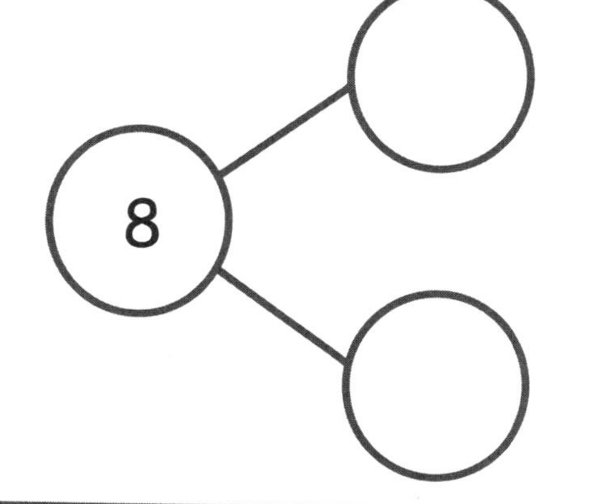

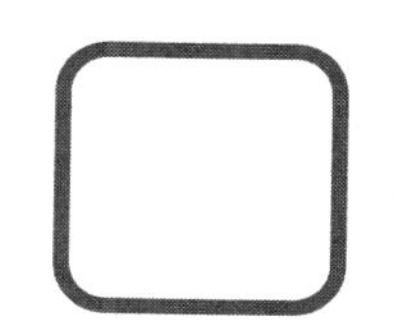

Draw 8 dots. Put an X on 1 dot.

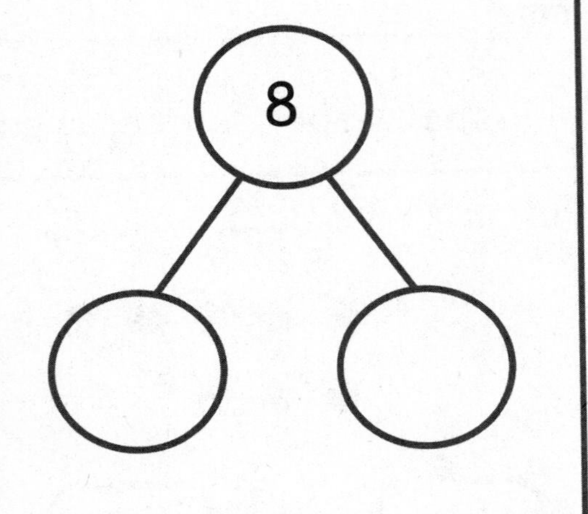

☐ - ☐ = ☐

Draw 8 dots in a 5-group. Put an X on 7 dots.

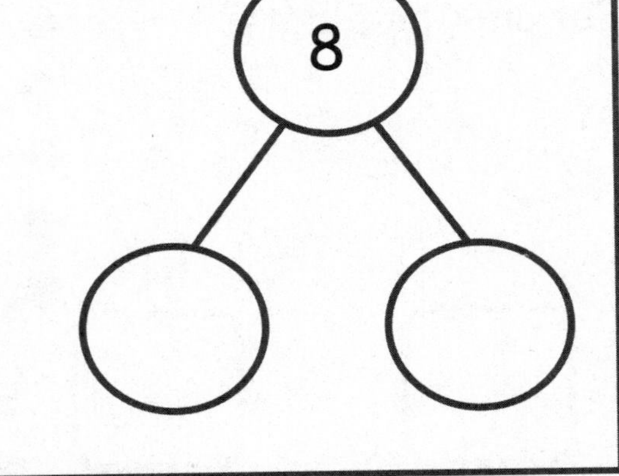

☐ - ☐ = ☐

Draw 8 dots in a 5-group. Put an X on some dots.

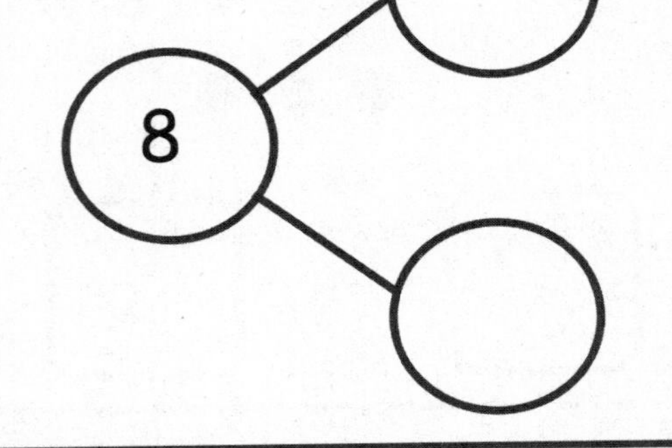

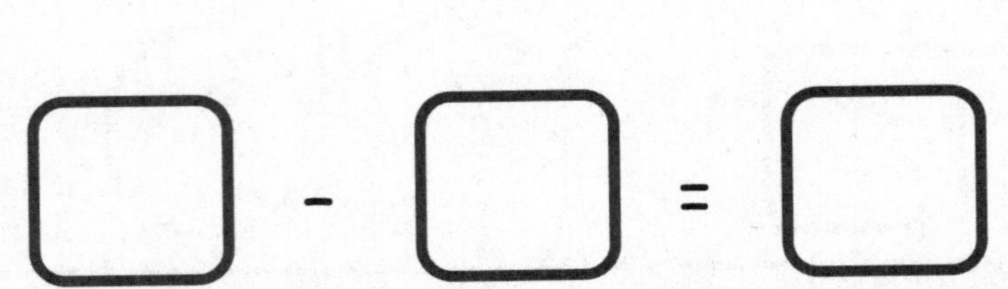

There were 9 flowers in Casey's garden.

She had 2 vases.

Draw 1 way she could have put all the flowers into the vases.

Draw

Show your picture to a friend. Did your friend draw the flowers in the vases the same way? Are both ways right? Are there other ways you could have shown the flowers?

Name ______________________________ Date ______________

There are 9 shirts. Color some with polka dots and the rest with stripes. Fill in the number bond.

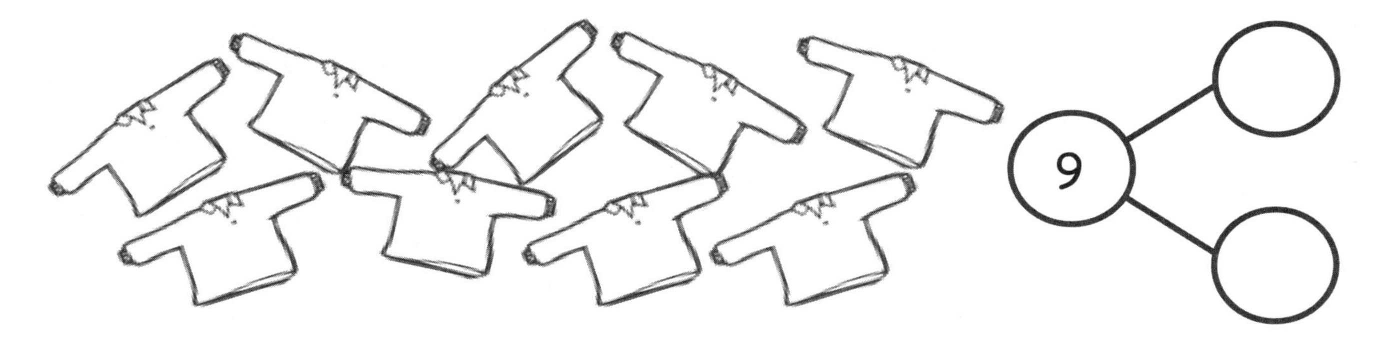

There are 9 flowers. Color some yellow and the rest red. Fill in the number bond.

There are 9 hats. Color some brown and the rest green. Fill in the number bond.

There are 9 jellyfish. Color some blue and the rest a different color. Fill in the number bond.

There are 9 butterflies. Color some butterflies orange and the rest a different color. Fill in the number bond.

Draw 9 balloons. Color some red and the rest blue. Make a number bond to match your drawing.

There are 9 socks.

Some are green, and the rest are blue.

Draw the set of green socks and the set of blue socks.

Draw

Make a number bond to show the green and blue socks. Show your picture to a friend. Do they look alike? How could you show the socks a different way?

Name ______________________ Date ____________

The squares below represent cube sticks.

Draw a line from the cube stick to the matching number bond. Fill in the number bond if it isn't complete.

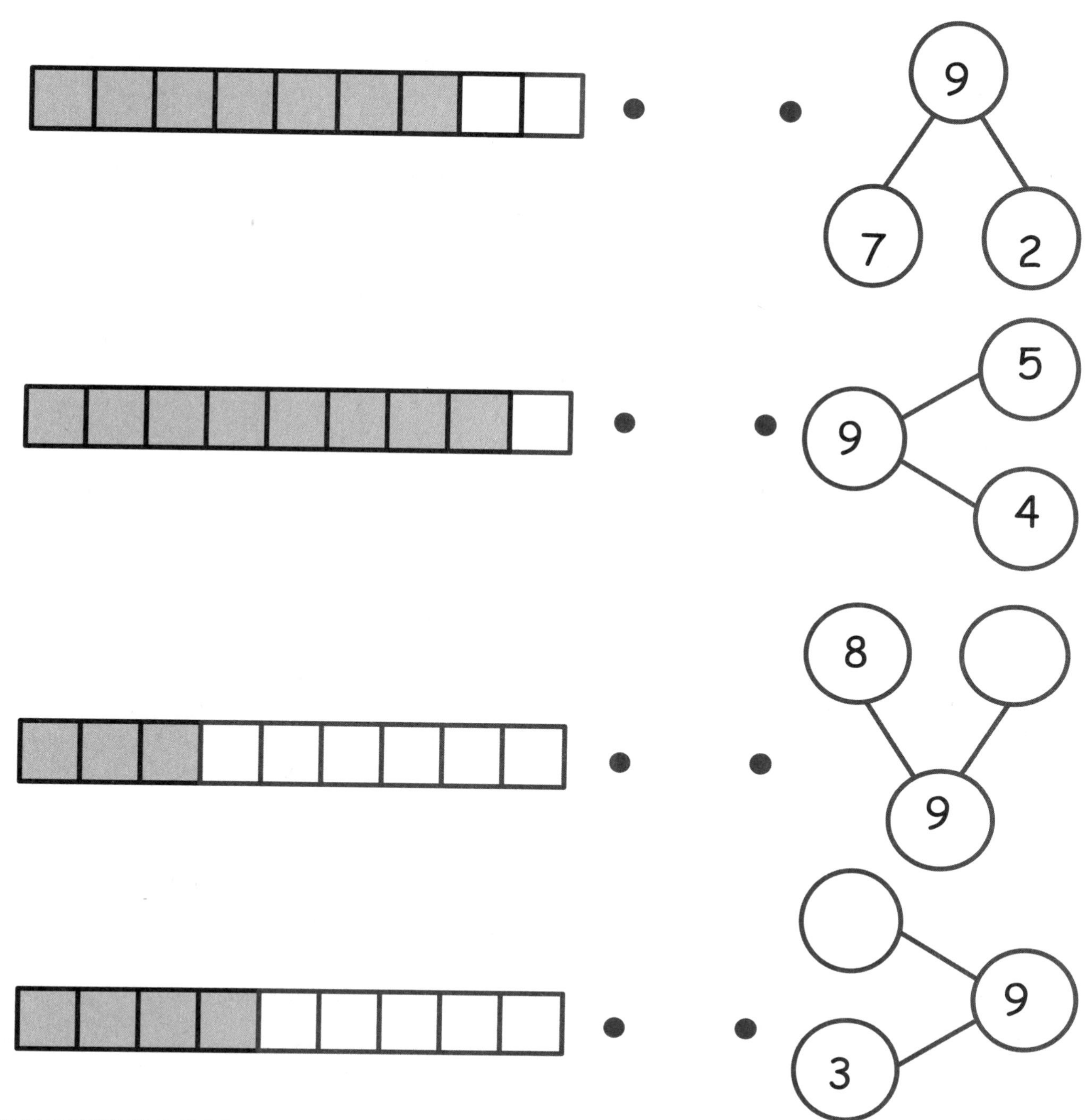

Draw and color cube sticks to match the number bonds. Fill in the number bond if it isn't complete.

9
5
4

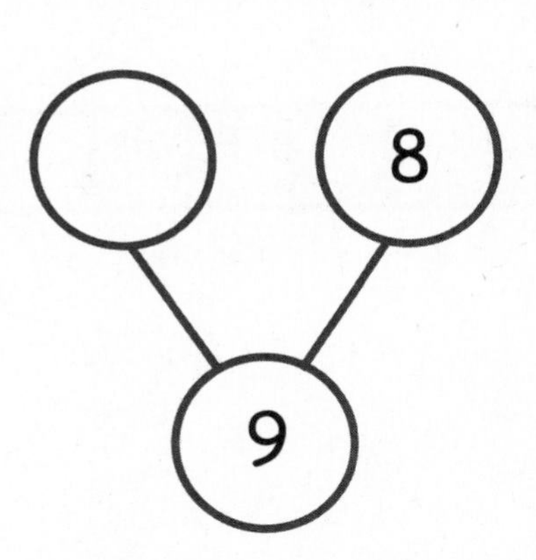

Create your own 9-cube stick, and fill in the number bond to match.

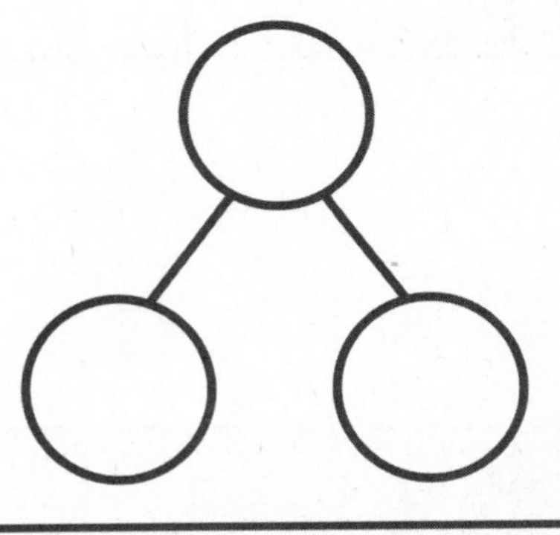

It's your birthday party!

You need 10 party hats for your friends.

Draw 10 hats.

Color some hats red and some blue.

Draw

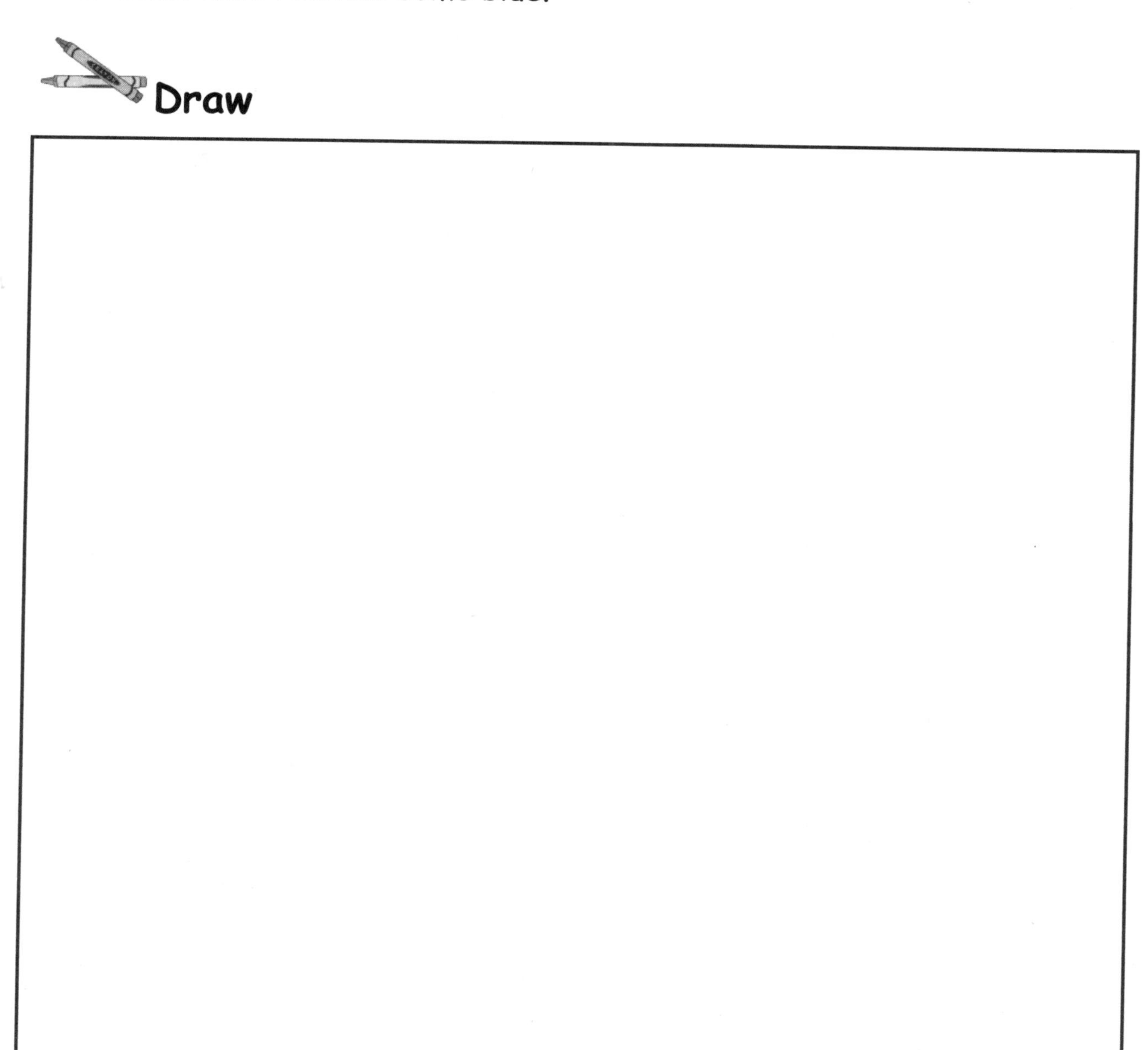

Make a number bond to go with your picture. Turn and talk with your partner. Do your pictures look the same? Explain to your partner how you decided which way to color your hats. Talk about how your number bonds are the same or different.

Name ______________________________ Date ______________

Benjamin had 10 bananas. He dropped some of the bananas. Fill in the number bond to show Benjamin's bananas.

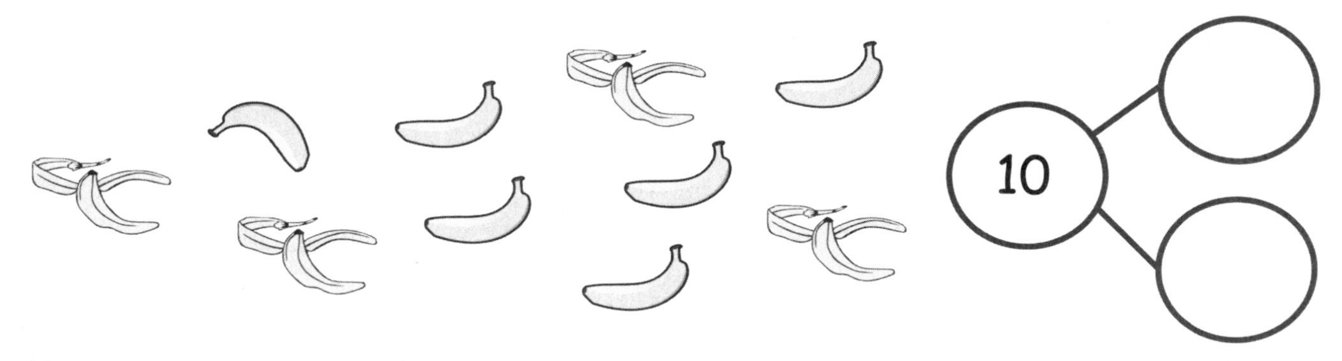

10

Savannah has 10 pairs of glasses. 5 are green, and the rest are purple. Color and fill in the number bond.

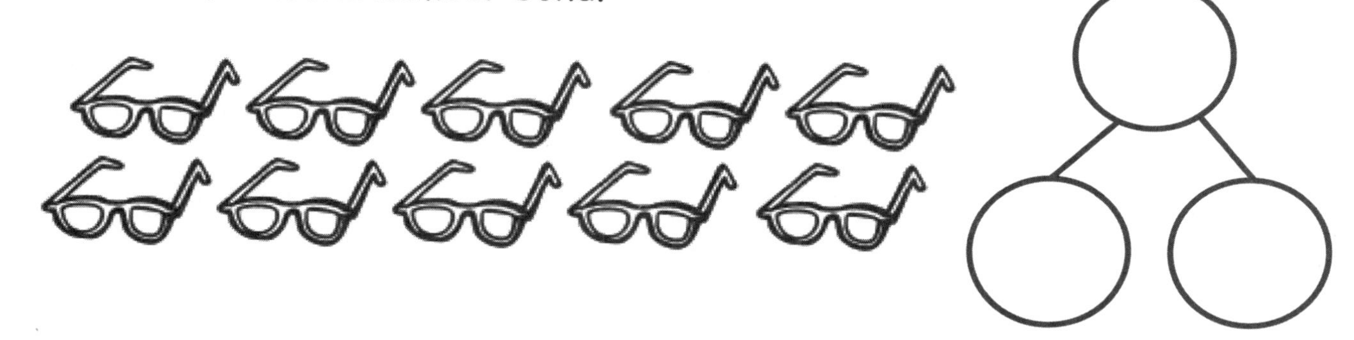

Xavier had 10 baseballs. Some were white, and the rest were gray. Draw the balls, and color to show how many may be white and gray. Fill in the number bond.

There were 10 dragons playing. Some were flying, and some were running. Draw the dragons. Fill in the number bond.

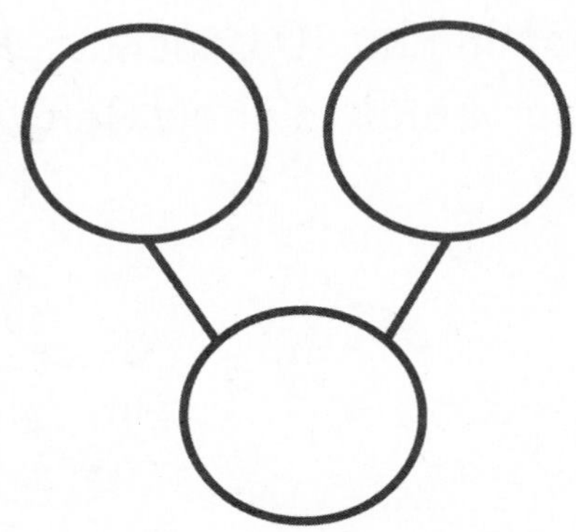

Create your own story of 10. Draw your story and a number bond to go with it.

Use your clay to make 10 tiny grapes.

Put some of the grapes on one plate and the rest on the other plate.

How many grapes do you have in all?

Draw

(Give students a ball of clay.) Draw a number bond to show the grapes. Talk about your grapes with a partner. Did she do it in the same way? Take off the grapes and try again.

Name ______________________________ Date ______________

These squares represent cube sticks. Look at the linking cube sticks. Draw a line from the cube stick to the number bond that matches. Fill in the number bond if it is not complete.

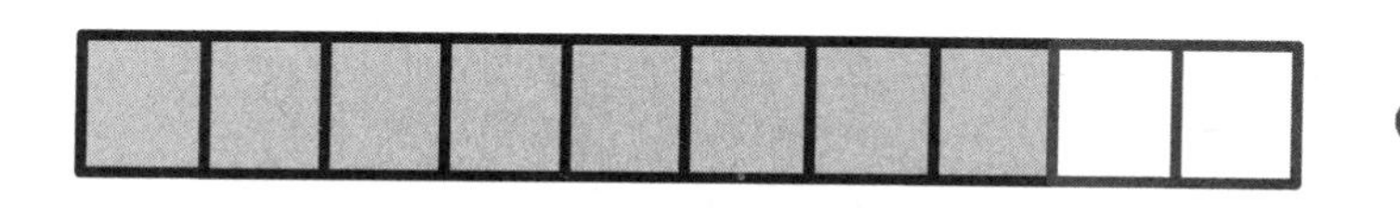

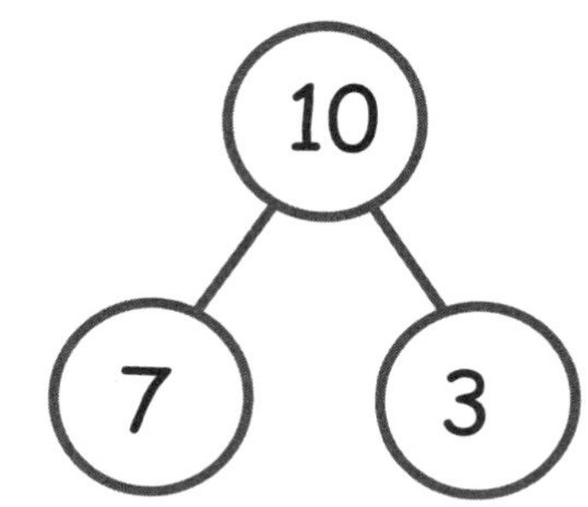

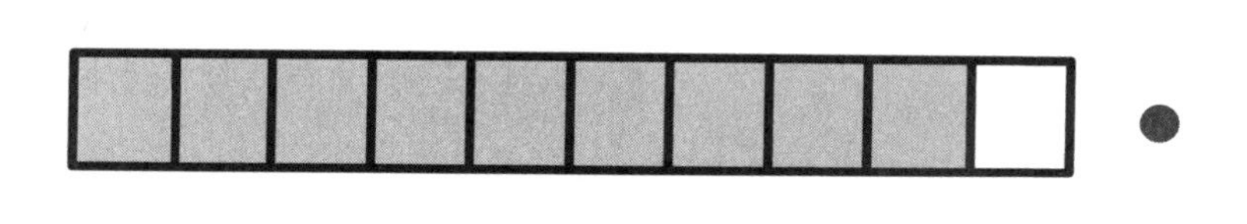

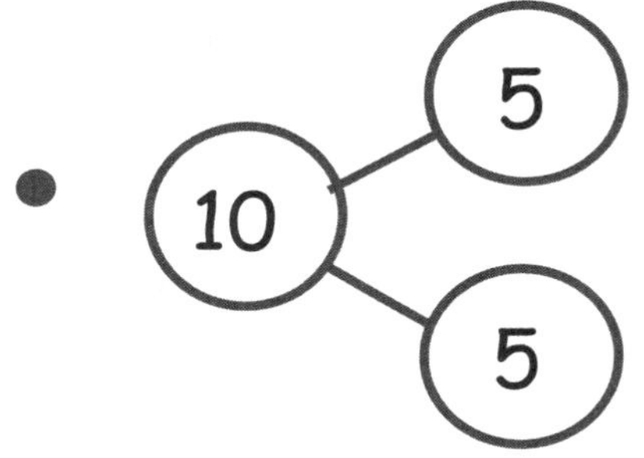

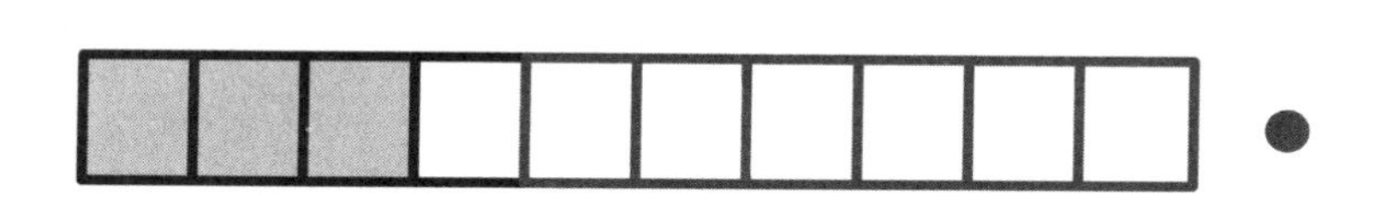

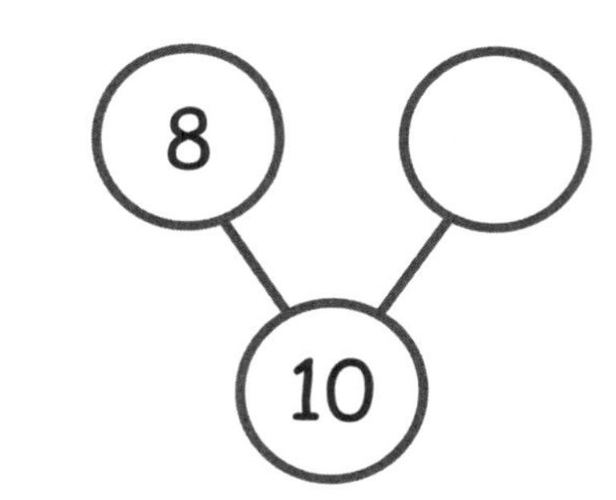

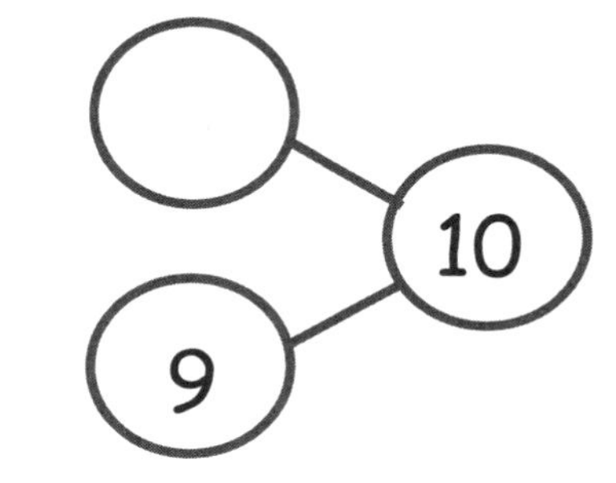

Draw and color cube sticks to match the number bonds.

10
5
5

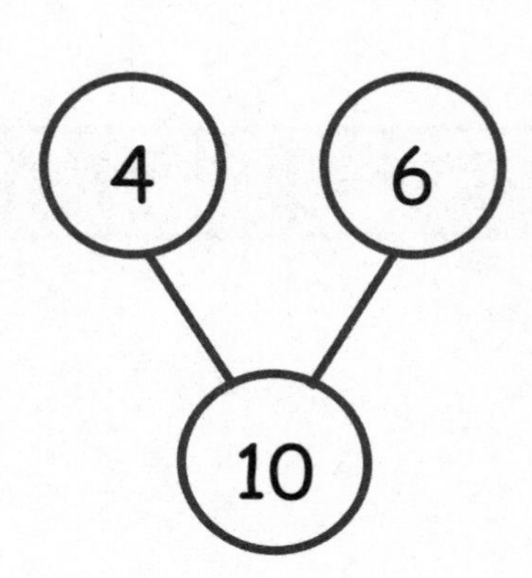

Create your own 10-cube stick, and fill in the number bond.

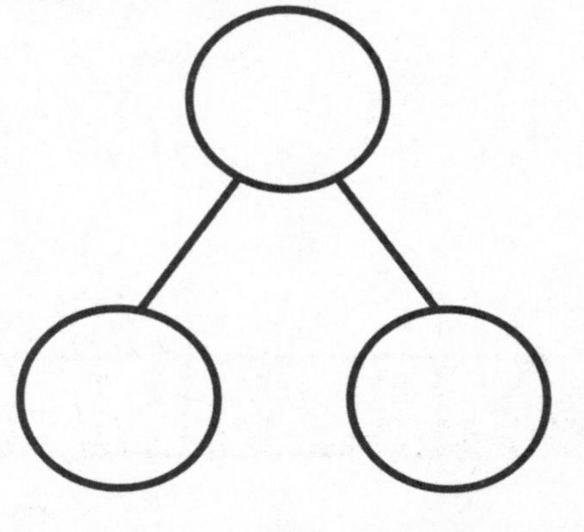

Emma had 9 pennies. Show her pennies in the middle of the desk.

She wanted to use 4 pennies to buy gum and 5 pennies to buy a balloon.

Count and slide apart the pennies she needs to buy gum and the balloon.

Draw

(Give students 9 pennies.) Draw a number bond to go with the story. Slide your groups of pennies together again. How many pennies in all? How could you create a new number bond with your pennies? Turn and talk to your partner about your work.

Name ______________________ Date ____________

Izzy had a tea party with 7 teddy bears and 2 dolls. There were 9 friends at the party. Fill in the number bond and number sentence.

9 = ☐ + ☐

Robin had 9 vegetables on her plate. She had 3 carrots and 6 peas. Draw the carrots and peas in the 5-group way. Fill in the number sentence.

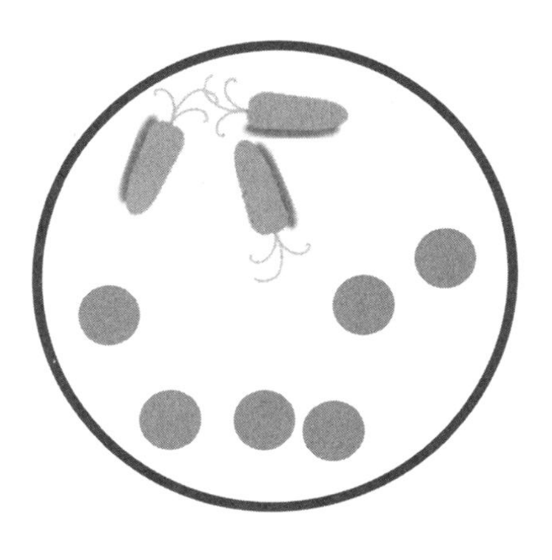

9 = ☐ + ☐

Shane played with 5 toy zebras and 4 toy lions. He had 9 animal toys in all. Draw black and tan circles to show the zebras and the lions in the 5-group way. Fill in the number sentence.

Jimmy had 9 marbles. 8 were red, and 1 was green. Draw the marbles in the 5-group way. Fill in the number bond and number sentence.

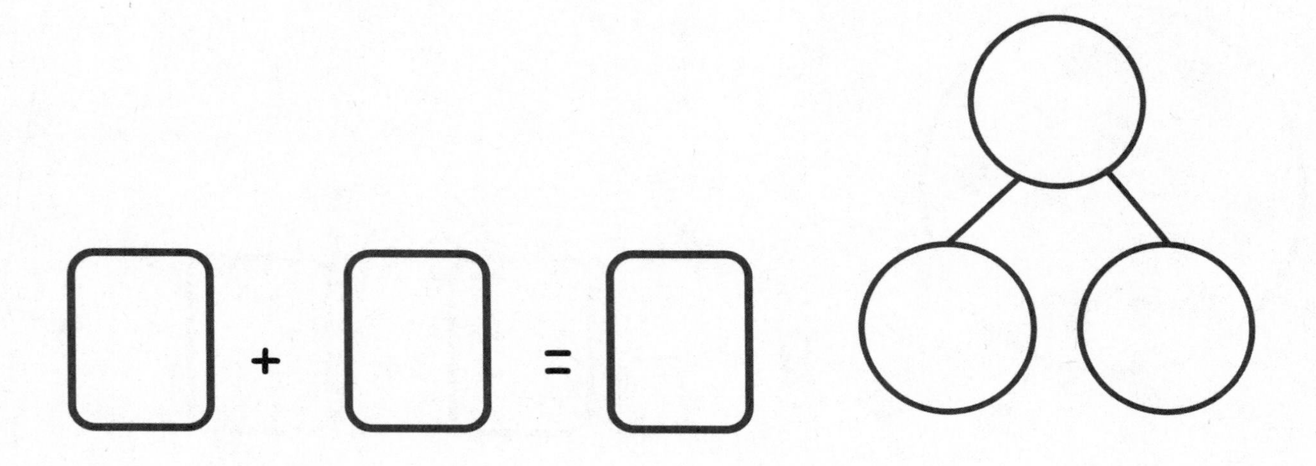

Pretend your linking cubes are pears from the pear tree.

Put 5 pears on the tree and 5 pears on the ground.

Draw

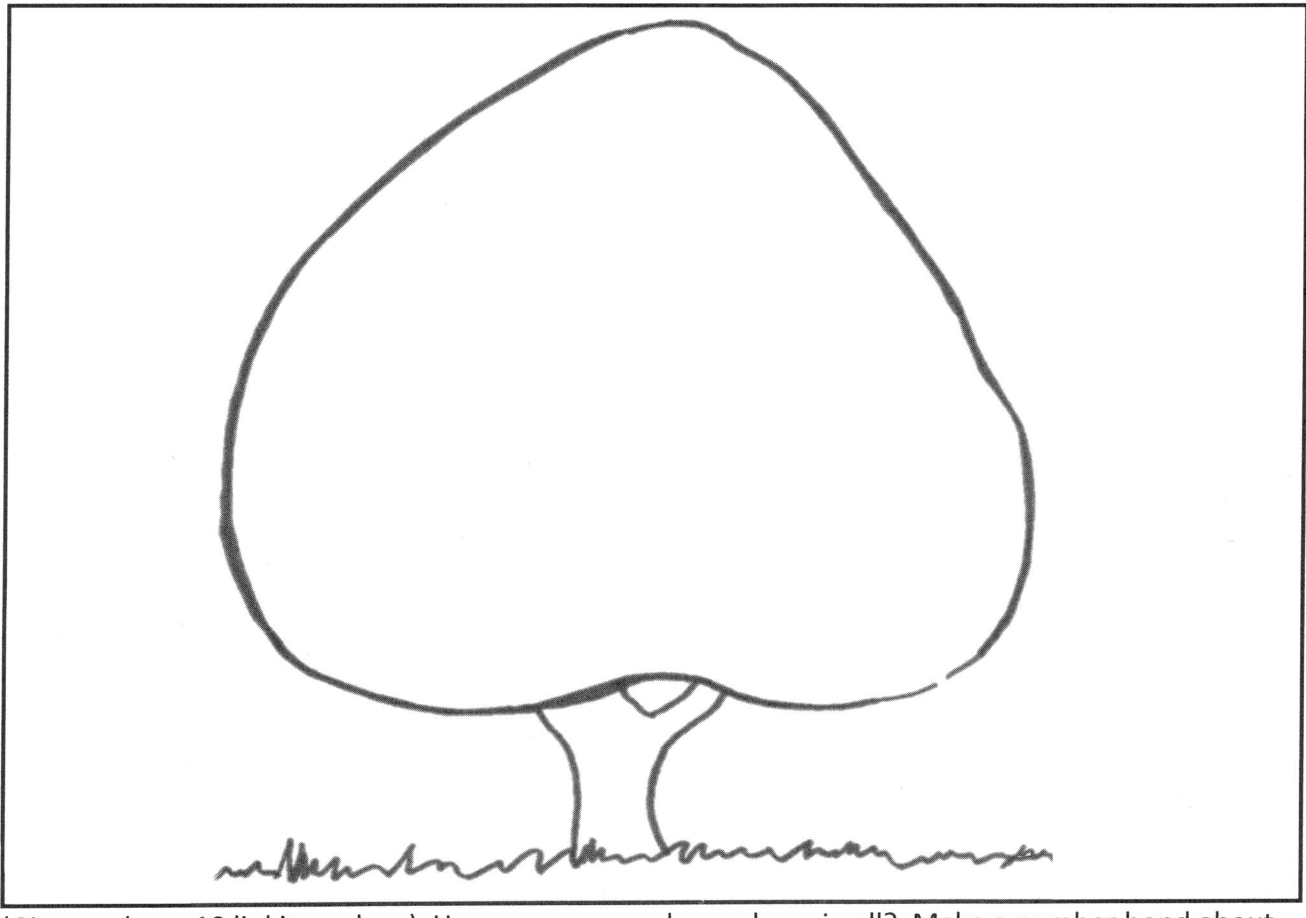

(Give students 10 linking cubes.) How many pears do you have in all? Make a number bond about the pears. (Number bonds can be written on the next page or on a personal white board.) Tell a partner about the pears. Can you think of a number sentence? Now, show another pear falling out of the tree. How many pears are in the tree now? Would your number bond change? Is there a different number sentence you would use to tell about what you did? Talk about your ideas with a partner.

Write

Name ______________________ Date ____________

Fill in the number bonds, and complete the number sentences.

Ricky has 10 space toys. He has 7 rockets and 3 astronauts.

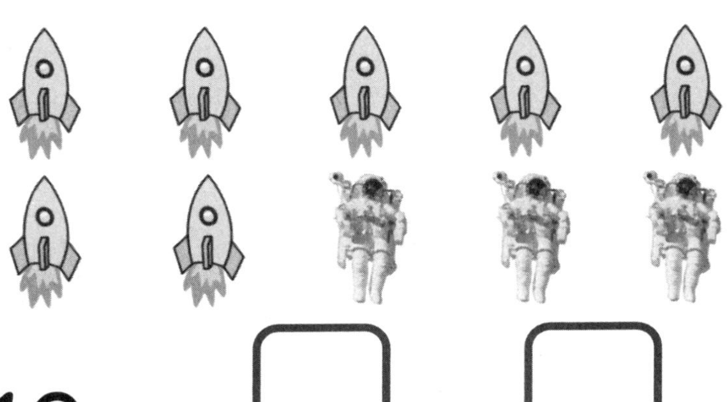

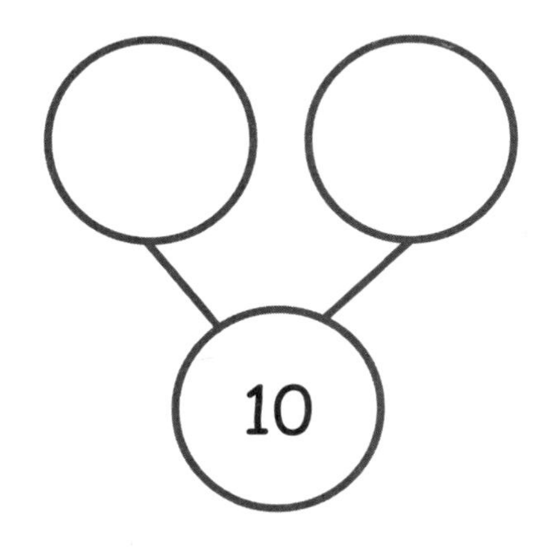

$10 = \square + \square$

Bianca has 4 pigs and 6 sheep on her farm. She has 10 animals altogether.

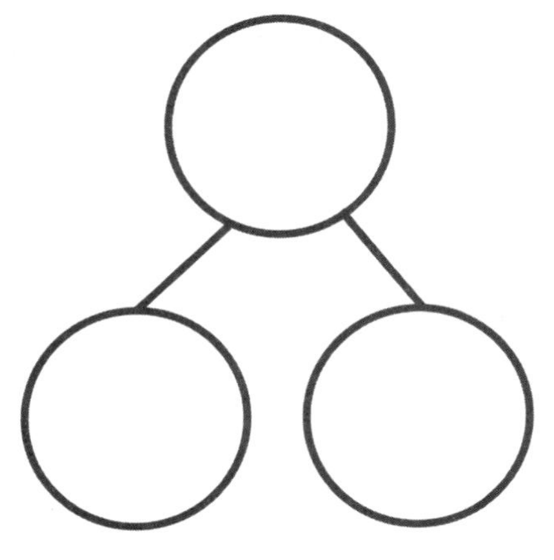

Danica had 5 green balloons. Her friend gave her 5 blue balloons. Draw all of her balloons in the 5-group way. Fill in both number sentences.

☐ = ☐ + ☐

☐ + ☐ = ☐

Jason is playing with 10 bouncy balls. He has 8 on the table and 2 on the floor. Draw the bouncy balls in the 5-group way. Fill in both number sentences.

☐ + ☐ = ☐

☐ = ☐ + ☐

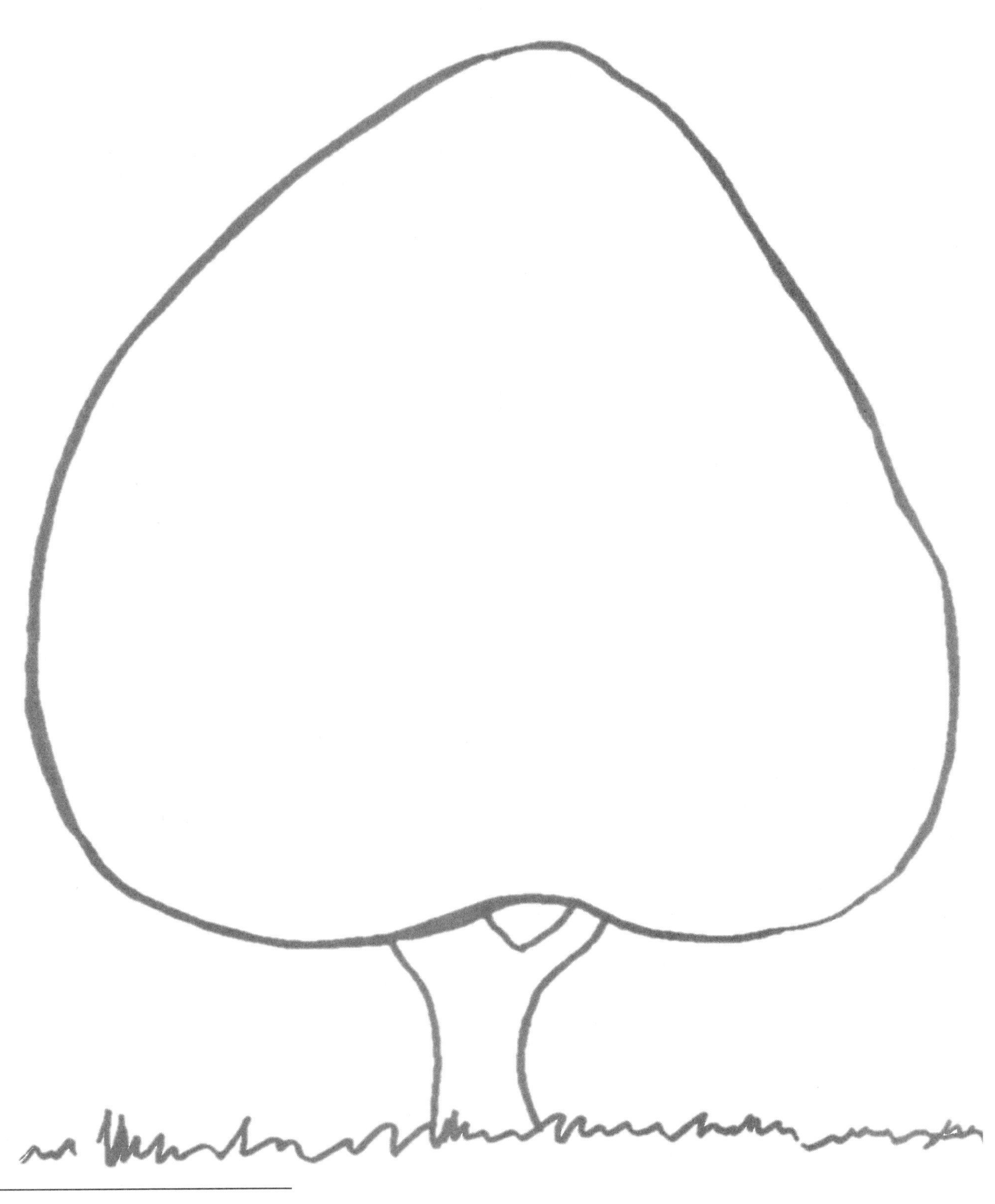

tree

5 children were playing soccer in the park.

Draw the children.

4 more children came to play.

Draw the new players.

Draw

How many children were playing soccer? How did you know? Turn and talk to your partner about your answer. Do you agree?

Name ____________________ Date ____________

Draw the story. Fill in the number sentence.

Zayne had 6 round crackers and 3 square crackers. How many crackers did Zayne have in all?

_____ + _____ = _____

Riley had 9 crayons. Her friend gave her 1 crayon. How many crayons did Riley have in all?

_____ + _____ = _____

Draw the story. Write a number sentence to match.

Jenny had 3 red and 7 purple pieces of construction paper. How many pieces of construction paper did Jenny have altogether?

Rhett had 5 square blocks. His friend gave him 4 rectangle blocks. How many blocks did Rhett have altogether?

equation

Lesson 31: Solve *add to with total unknown* and *put together with total unknown* problems with totals of 9 and 10.

135

Chen had 9 pencils.

Some of his pencils were red, and some were blue.

Draw Chen's pencils.

Draw

Make a number bond to show the pencils. Turn and talk to your partner about your picture and number bond. Do your pictures look the same? Are your number bonds the same? Are they both correct?

Name ______________________________ Date ______________

Listen to the word problem. Fill in the number sentence.

Cecilia has 9 bows. Some have polka dots, and some have stripes. How many polka dot and how many striped bows do you think Cecilia has?

9 = ☐ + ☐

Keegan has 10 train cars. Some are black, and some are green. How many black and green train cars do you think Keegan has?

10 = ☐ + ☐

Kate has 9 heart stickers. Some are yellow, and the rest are green. Show two different ways Kate's stickers could look. Fill in the number sentences to match.

9 = ☐ + ☐

9 = ☐ + ☐

Danny has 10 robots. Some are red, and the rest are gray. Show two different ways Danny's robots could look. Fill in the number sentences to match.

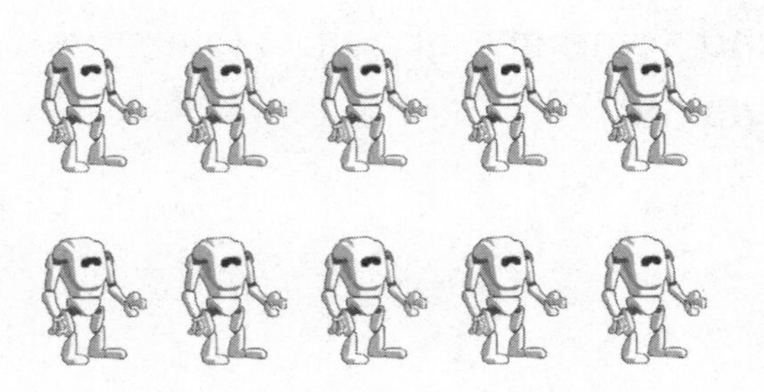

10 = ☐ + ☐

10 = ☐ + ☐

Name ______________________________ Date ______________

Color the robots to match the number sentence. Tell a story about the robots.

10 = 5 + 5

10 = 6 + 4

10 = 7 + 3

10 = 8 + 2

10 = 9 + 1

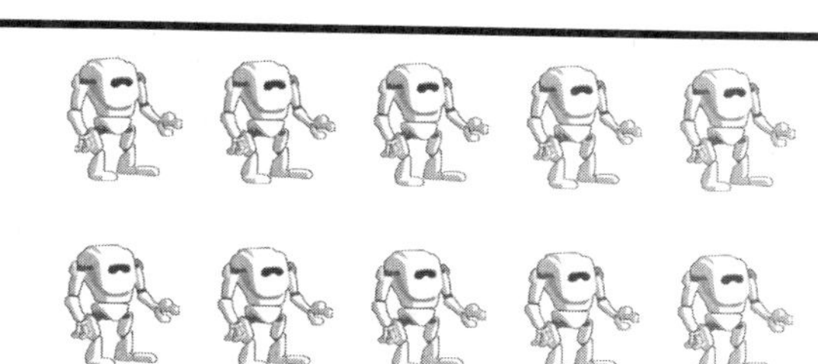

Pretend your linking cubes are ants and the box is a picnic blanket.

Put all the ants on the blanket.

Now, pretend some of the ants crawled off the blanket.

Slide some ants off the blanket to show the ones that crawled away.

Draw

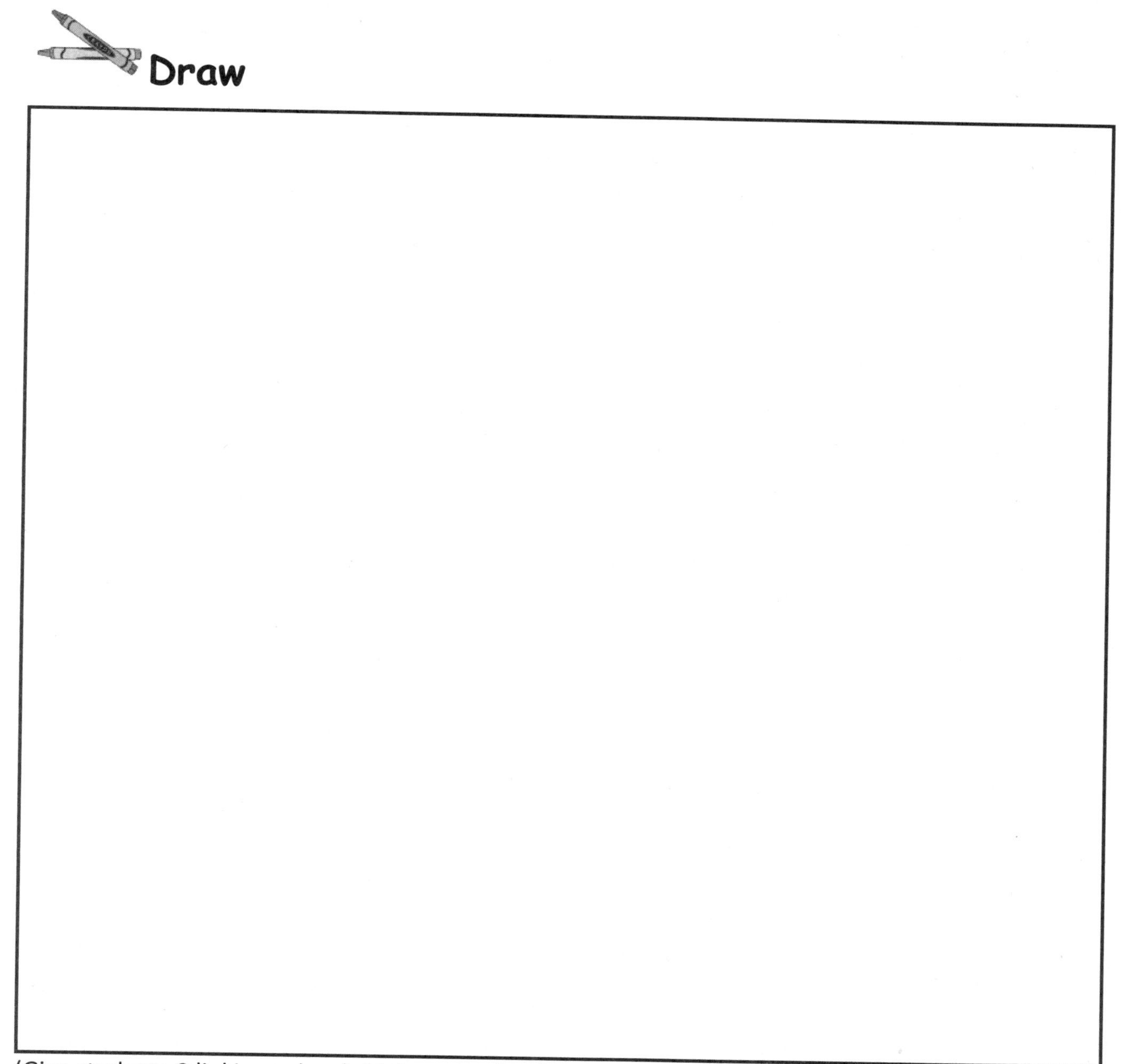

(Give students 9 linking cubes.) Make a number bond to show the ants that stayed on the blanket and the ants that crawled away. Show your number bond to a friend.

Name ______________________________ Date ______________

Fill in the number sentence to match the story.

There were 7 trains. 2 trains rolled away. Now there are 5 trains.

____ − ____ = ____

There were 9 cars at the stop sign. 7 drove away. There are 2 cars left.

____ − ____ = ____

There were 10 people. 6 people got on the bus. Now there are 4 people.

____ − ____ = ____

Draw the story. Fill in the number sentence to match.

The bus had 10 people. 5 people got off. Now there are 5 people left.

____ – ____ = ____

There were 9 planes in the sky. 3 planes landed. Now there are 6 planes in the sky.

____ – ____ = ____

subtraction equation

Tony had 8 checkers.

His friend took 3 away.

How many checkers did Tony have left?

Draw

Draw a picture of the story. Make a number bond and a number sentence about the story. Show your work to a friend. Did you both do it the same way?

Write

Name ____________________ Date ____________

Fill in the number sentences and number bonds.

There are 9 babies playing. 2 crawl away. How many babies are left?

9 - 2 = ____

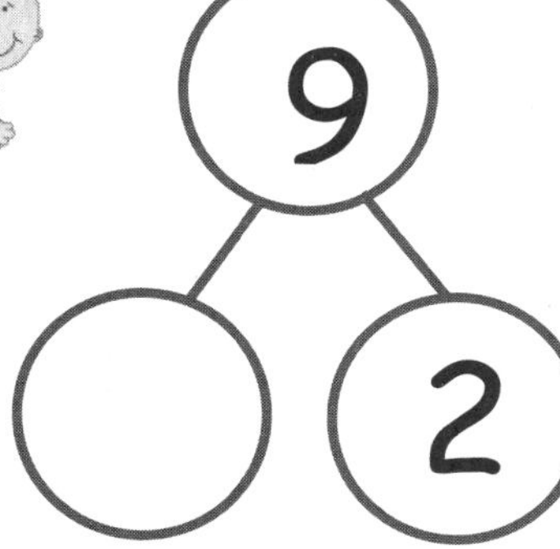

There are 10 babies playing. 1 crawls away. How many babies are left?

10 - ____ = ____

There are 9 babies playing. 6 crawl away. How many babies are left?

____ - ____ = ____

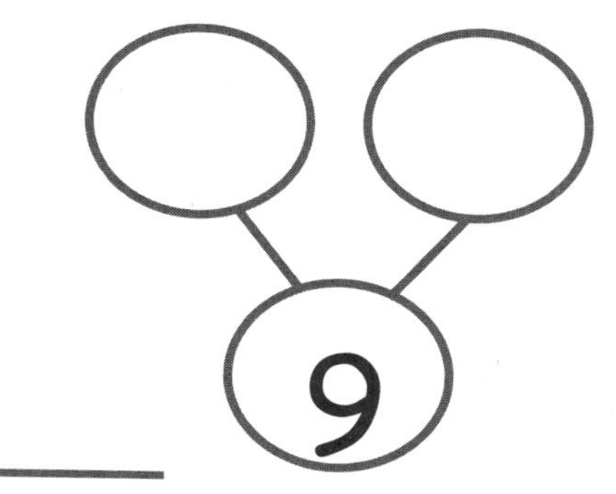

The squares below represent cube sticks.
Carlos had a 9-stick. He broke off 4 cubes to share with his friend.
How many cubes are left? Draw a line to show where he broke his stick.

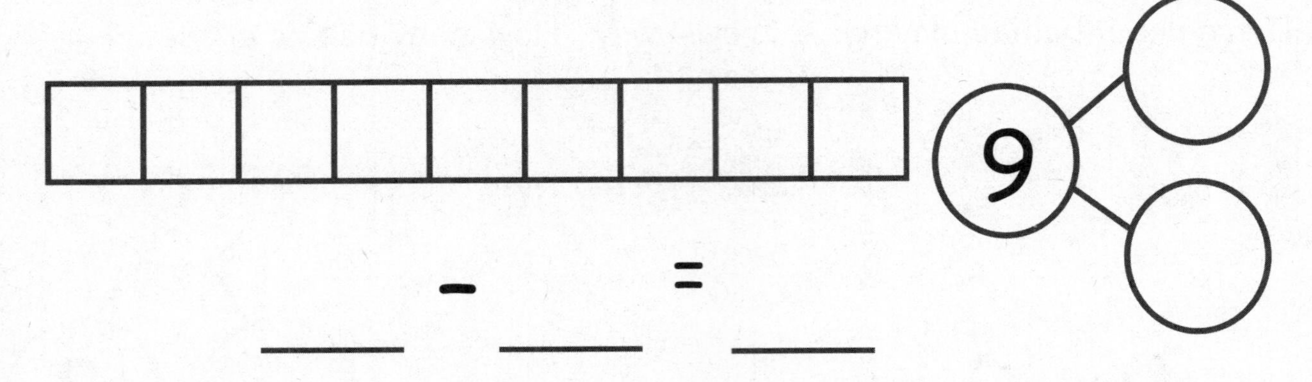

_____ − _____ = _____

Sophie had 10 grapes. She ate 6 grapes. How many grapes are left?
Draw her grapes, and cross off the ones she ate.

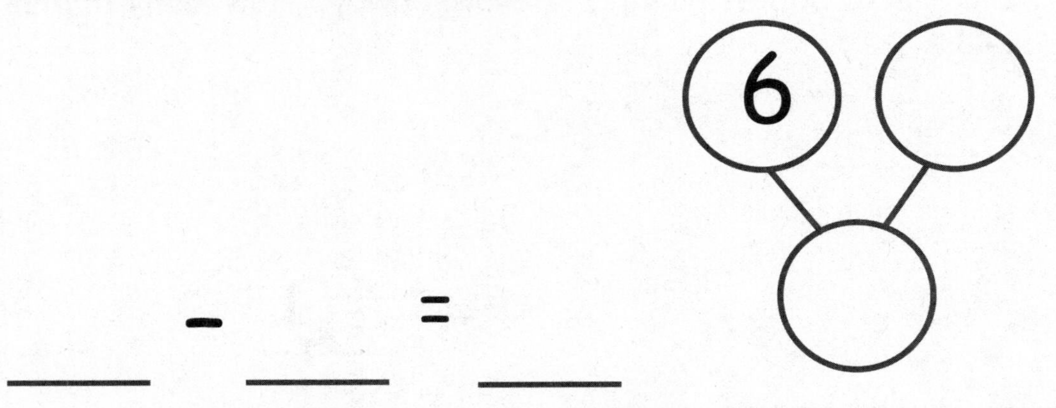

_____ − _____ = _____

Spot had 10 bones. He hid 8 bones in the ground. How many bones does he have now? Draw Spot's bones.

10

_____ − _____ = _____

Steve had 9 pennies.

He wanted to put some pennies into each of his two pockets.

Draw

(Give students 9 pennies.) Use your pennies to show one way he could have separated them. Make a number bond about your idea. Show your number bond to your partner. Did he do it the same way? How many different ways can you separate the pennies?

Name ______________________ Date ____________

Cross off the part that goes away. Fill in the number bond and number sentence.

Jeremy had 9 baseballs. He took 5 baseballs outside to play, and they got lost. How many balls are left?

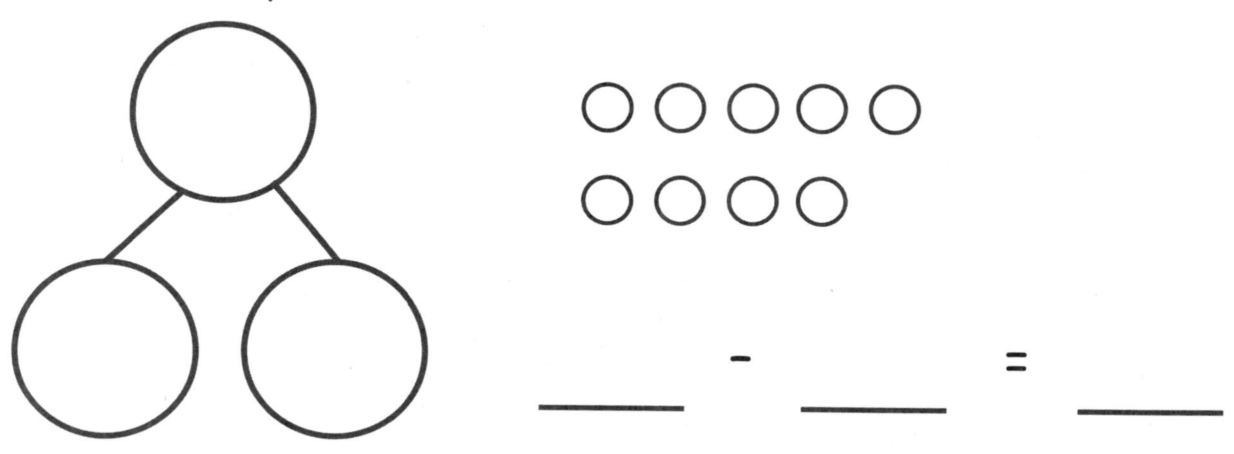

Sandy had 9 leaves. Then, 4 leaves blew away. How many leaves are left?

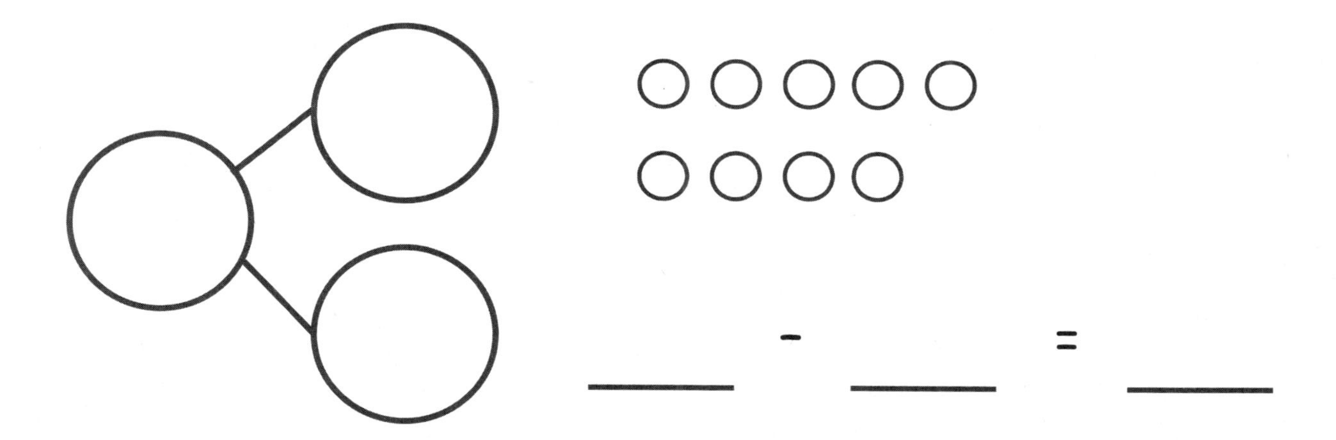

Make a 5-group drawing to show the story. Cross off the part that goes away. Fill in the number bond and number sentence.

Ryder had 9 star stickers. He gave 3 to his friend. How many star stickers does Ryder have now?

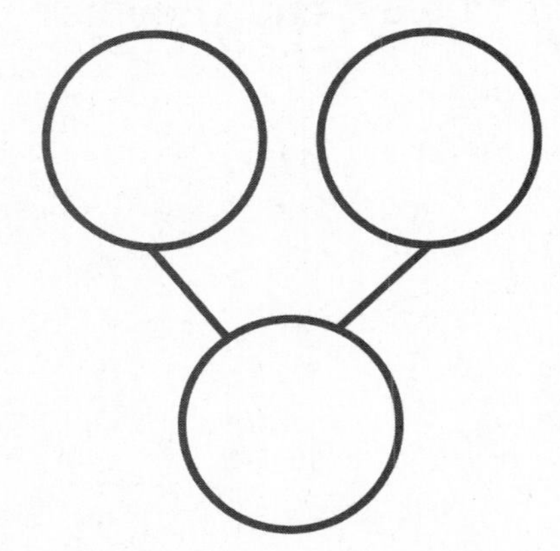

_____ - _____ = _____

Jen had 9 granola bars. She gave 8 of the granola bars to her teammates. How many granola bars does she have left?

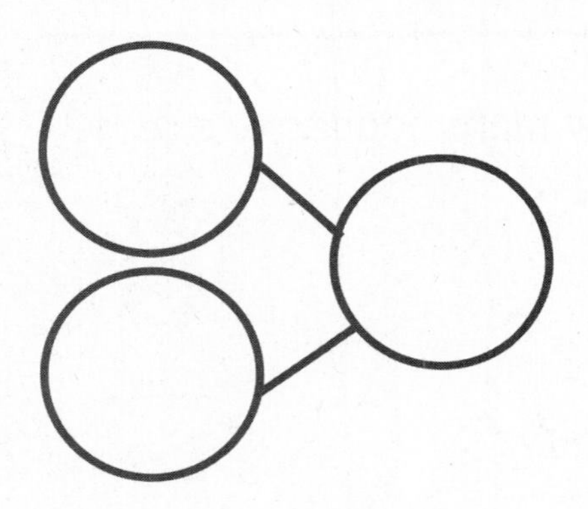

_____ - _____ = _____

Subtract.

2 - 1 = ☐ 3 - 2 = ☐ 4 - 3 = ☐ 5 - 4 = ☐

Martin had 10 building blocks.

Pretend your linking cubes are his blocks.

Count to make sure there are 10.

Draw

(Give students 10 linking cubes.) He shared 4 blocks with his sister. Move 4 blocks to show the ones he shared. How many blocks did he still have? Make a number bond about the story. Now, make a number sentence. Show your work to your partner. Did she do it the same way? Put your blocks back together. Act out the story again, sharing a different number of blocks this time. How does your number sentence change?

Name ______________________ Date ____________

Fill in the number bond and number sentence. Cross off the part that goes away.

Stan had 10 blueberries. He ate 5 berries. How many blueberries are left?

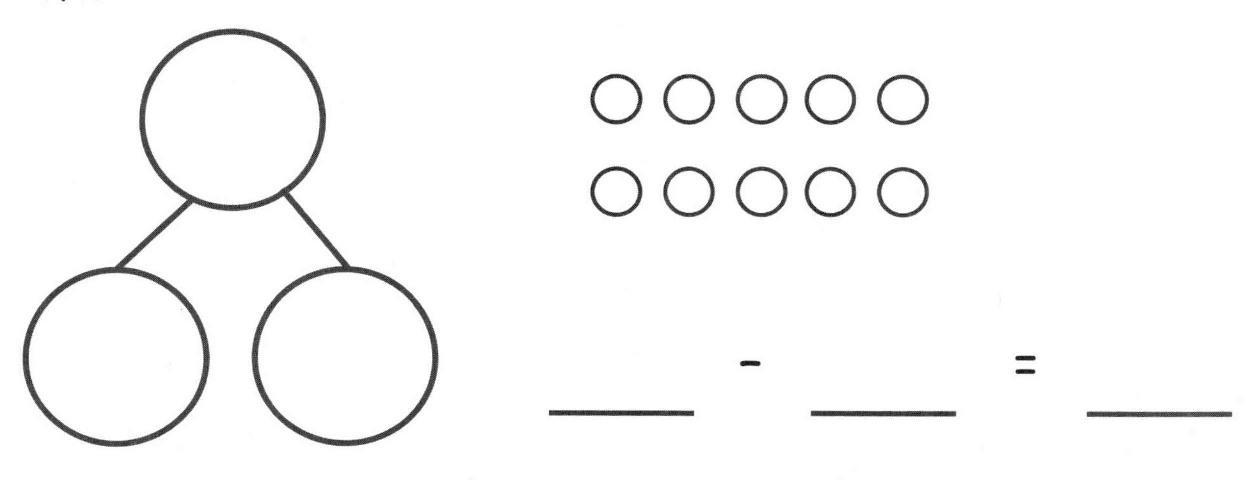

Tracy had 10 heart stickers. She lost 1 sticker. How many stickers are left?

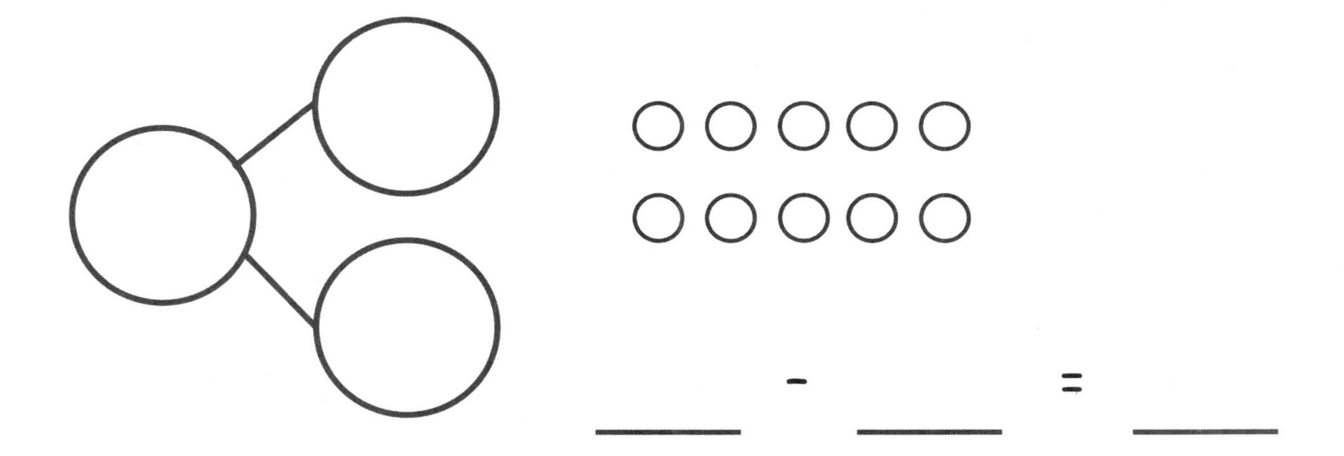

Make a 5-group drawing to show the story. Fill in the number bond and number sentence. Cross off the part that goes away.

Nick had 10 party hats. 7 hats were thrown away. How many hats does Nick have now?

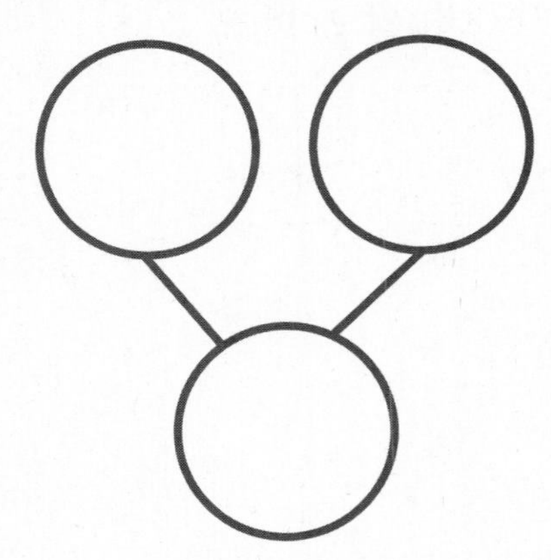

_____ - _____ = _____

Tatiana had 10 juice boxes. 3 juice boxes broke and spilled. How many full juice boxes does she have left?

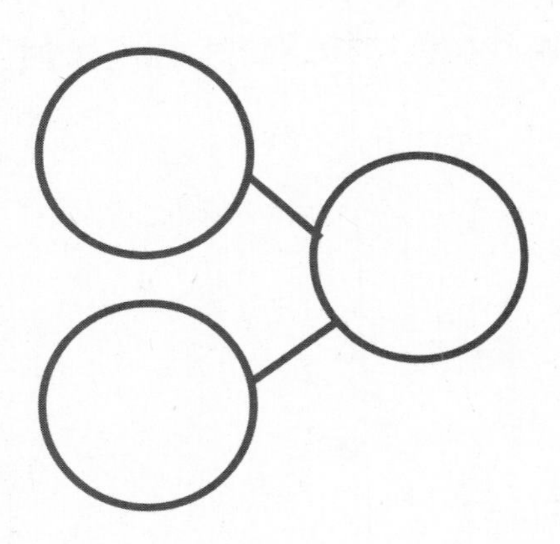

_____ - _____ = _____

Subtract.

5 - 1 = ☐ 5 - 2 = ☐ 5 - 3 = ☐ 5 - 4 = ☐

Chico the puppy had 8 tennis balls.

His owner threw 2 of the balls, but Chico brought them back.

Make 8 balls with your clay.

Draw

(Give students a ball of clay.) Show the story with the clay balls you created. Don't throw them. Remember, Chico brought them right back. Did Chico lose any of the tennis balls? Did he find any more? How many tennis balls does Chico have at the end of the story? Turn and tell your partner how you can create number sentences about Chico's adventures. Then, act out the story with different numbers.

Name ________________________ Date ______________

Listen to each story. Show the story with your fingers on the number path. Then, fill in the number sentence.

1	2	3	4	5	6	7	8	9	10

Freddy had 3 strawberries for a snack. His dad gave him 2 more strawberries. How many strawberries does Freddy have now?

3 + 2 = ____

Freddy ate 2 of his strawberries. How many strawberries does Freddy have now?

5 - 2 = ____

Logan had 7 frogs. 2 frogs hopped away. How many frogs does Logan have now?

____ - ____ = ____

Pretend that Logan's 2 frogs hopped back. How many frogs does he have now?

____ + ____ = ____

1	2	3	4	5	6	7	8	9	10

Stella had 4 pennies. She found 3 more pennies. How many pennies does Stella have now?

Stella gave the 3 pennies to her dad. How many pennies does she have now?

_____ – _____ = _____

Marissa made 8 bracelets. She loved them so much she did not give any away. How many bracelets does Marissa have now?

_____ – _____ = _____

Jackson found 6 toys under his bed. He looked and did not find any more toys. How many toys does Jackson have now?

_____ + _____ = _____

Solve.

2 + 0 = ☐ 2 – 0 = ☐ 4 – 0 = ☐ 3 + 0 = ☐

Pretend your cubes are dinosaurs.

1 dinosaur went to the watering hole because he was thirsty.

Move 1 of your cubes to the watering hole.

1 more dinosaur got thirsty, too.

Add another cube to the watering hole.

How many dinosaurs are at the watering hole?

Draw

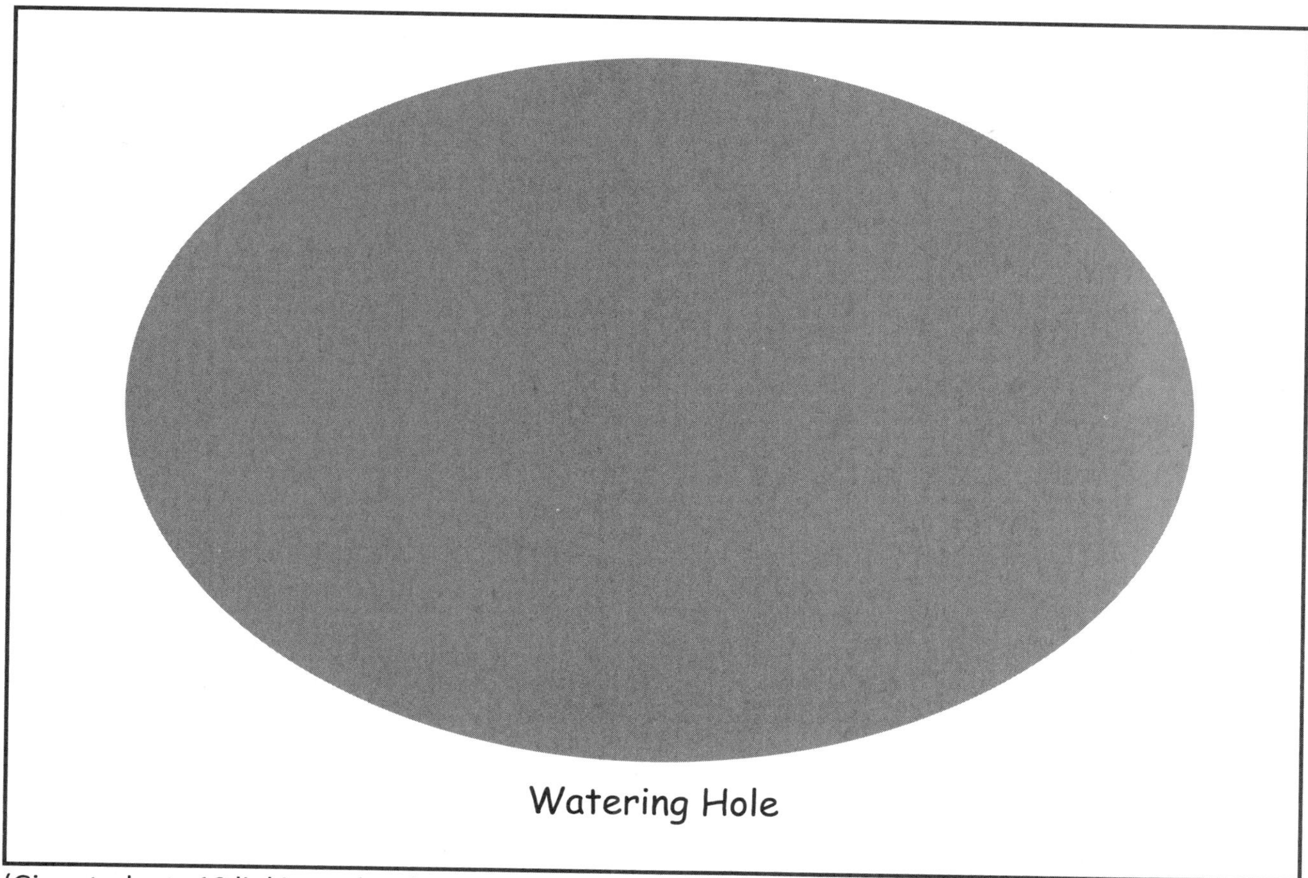

(Give students 10 linking cubes.) Turn to a partner, and talk about an addition sentence that would tell what you just did. Another dinosaur got thirsty. Add another cube to the watering hole. Now how many dinosaurs are at the watering hole? Talk to your partner about the new addition sentence. Keep acting out the story until all the dinosaurs are drinking water. Do you notice any patterns?

Name ______________________ Date ____________

1	2	3	4	5	6	7	8	9	10

Use the number path to add. Write the number in the box. Color the circles to match. Use a different color to show 1 more.

$1 + 1 = \square$

$2 + 1 = \square$

$3 + 1 = \square$

$4 + 1 = \square$

$5 + 1 = \square$

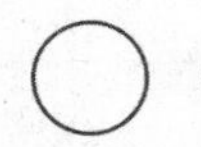
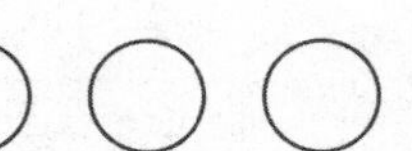

6 + 1 = ☐

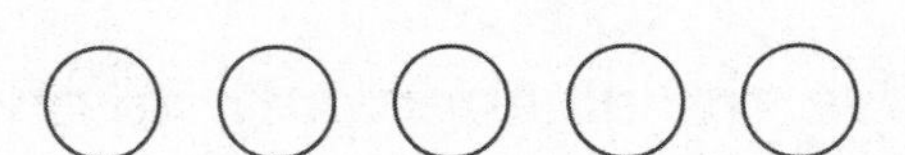

7 + 1 = ☐

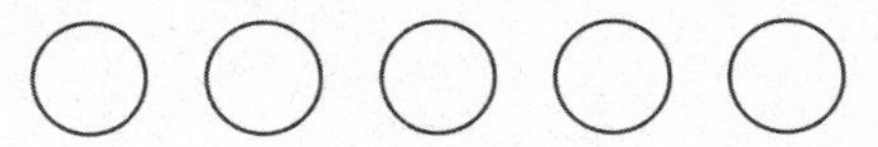

8 + 1 = ☐

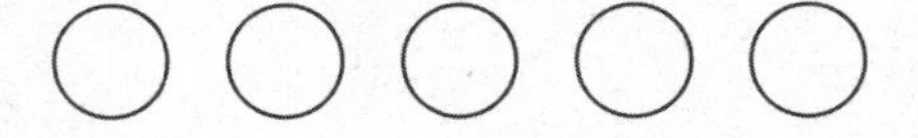

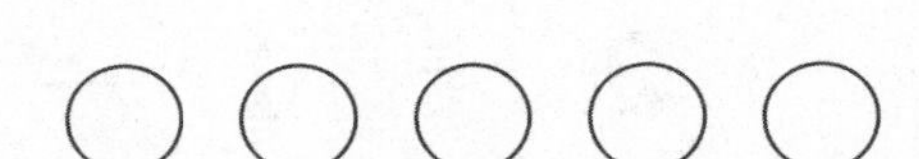

9 + 1 = ☐

Fill in the number sentences. Color the circles.

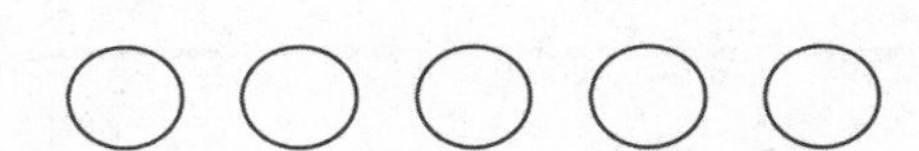

☐ + 1 = ☐

☐ + 1 = ☐

Tim had 10 friends. Draw his 10 friends.

Tim had 7 oranges. Draw his 7 oranges.

He wanted to give an orange to each of his friends.

Does he have enough?

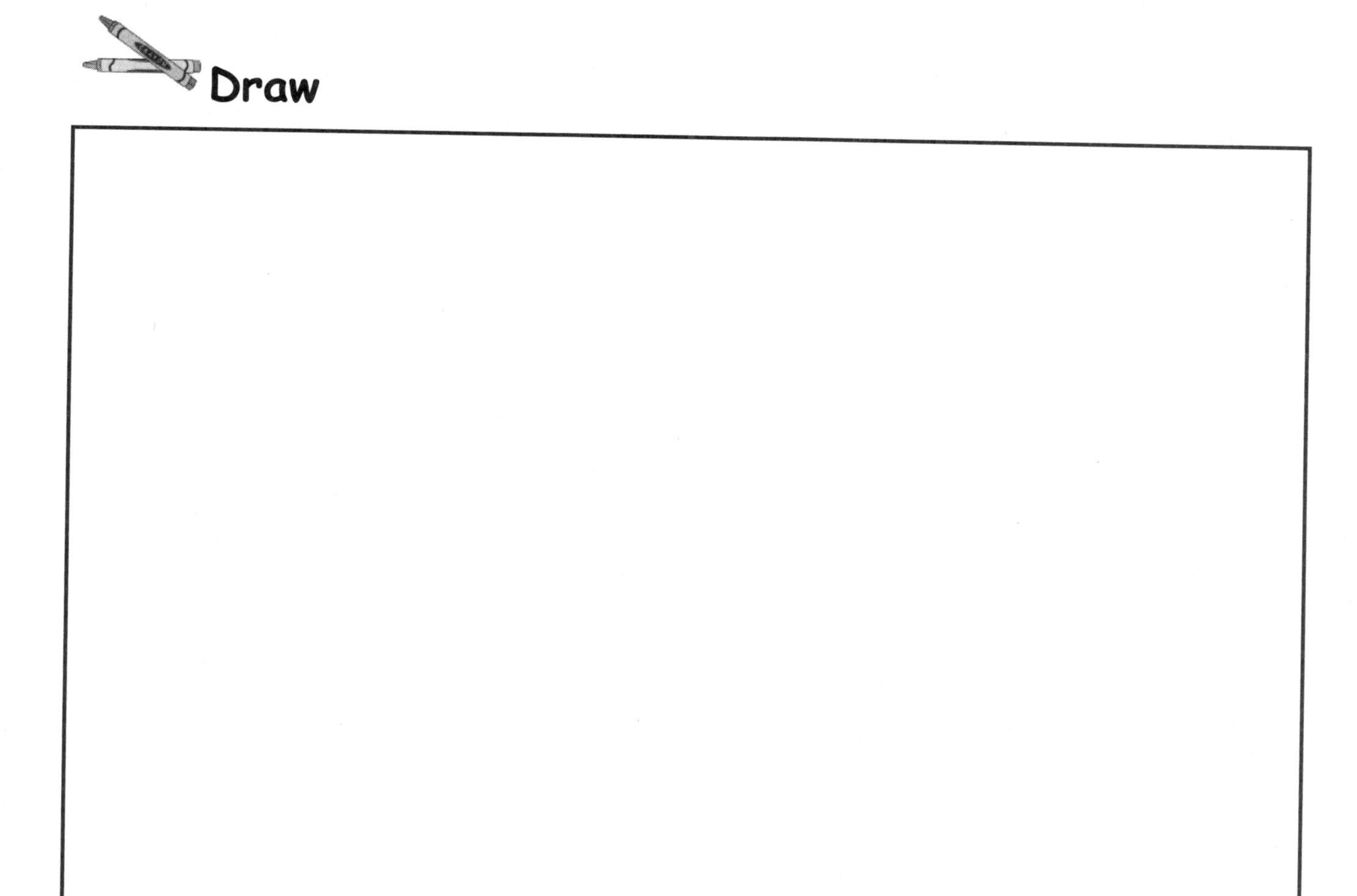

Draw more oranges so there are enough for all of his friends. Circle the new oranges. How many more oranges did he need? Check your work by drawing a line to match each friend with an orange. Now, show your work to a friend. Did she do it the same way? Talk about what would have happened if Tim had started with 8 oranges.

Name ______________________ Date ____________

Draw dots to make 10. Fill in the number bond.

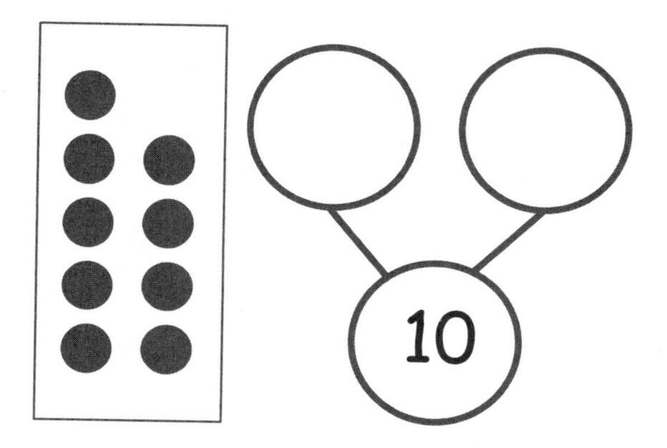

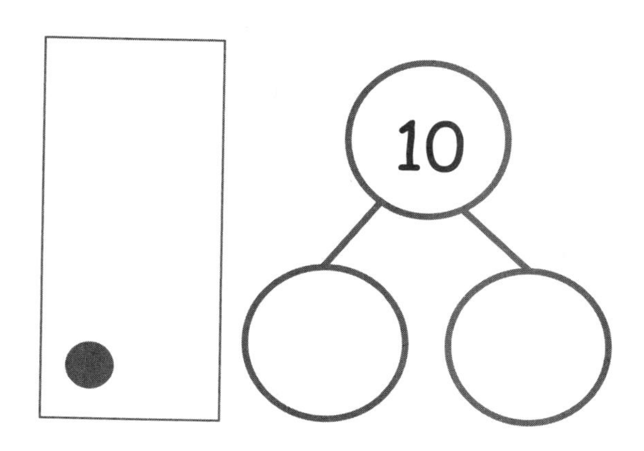

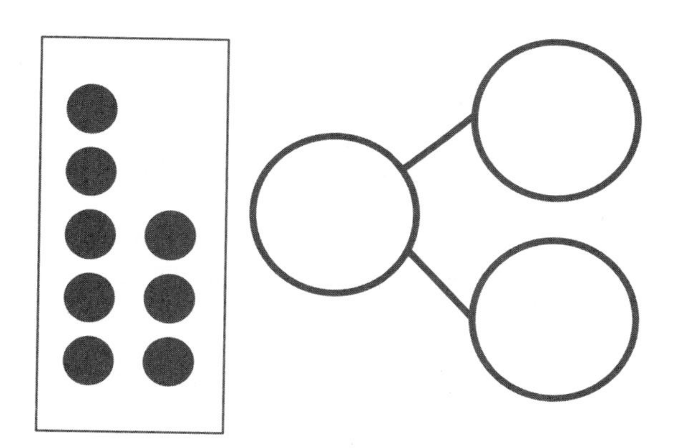

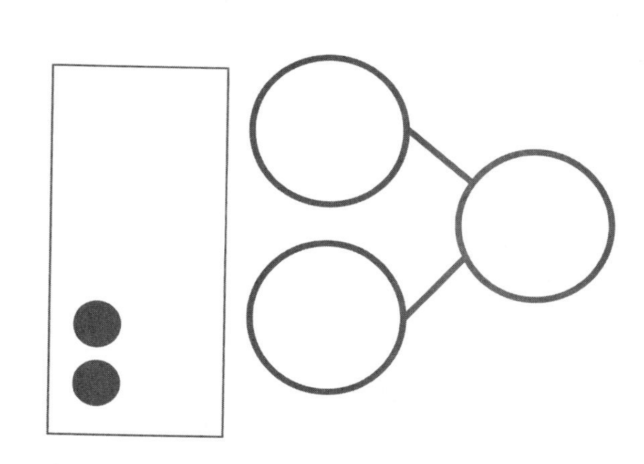

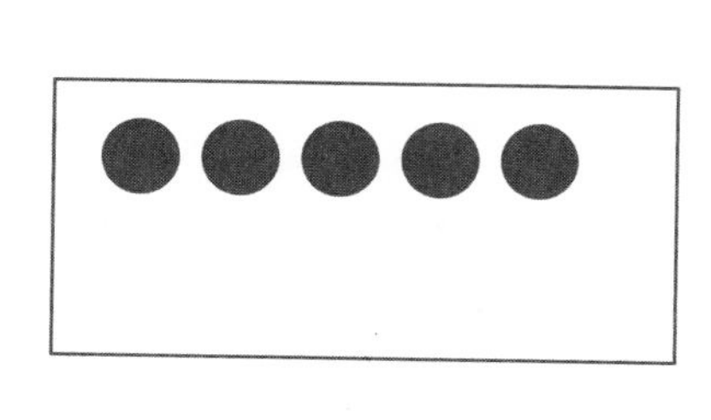

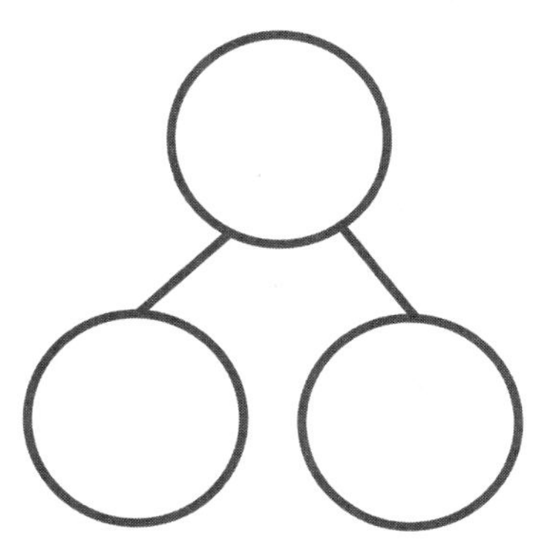

Draw dots to make 10. Fill in the number bond.

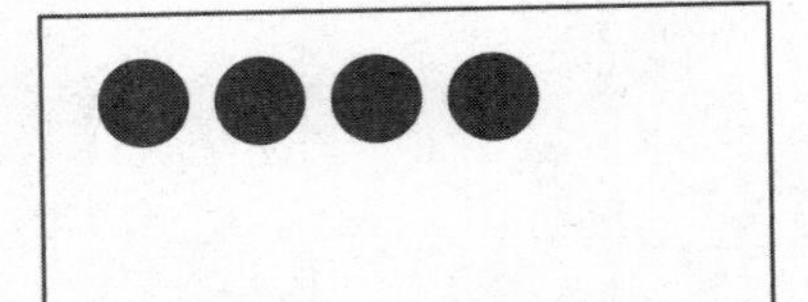

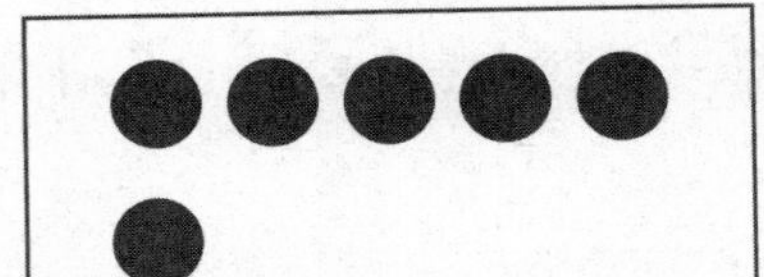

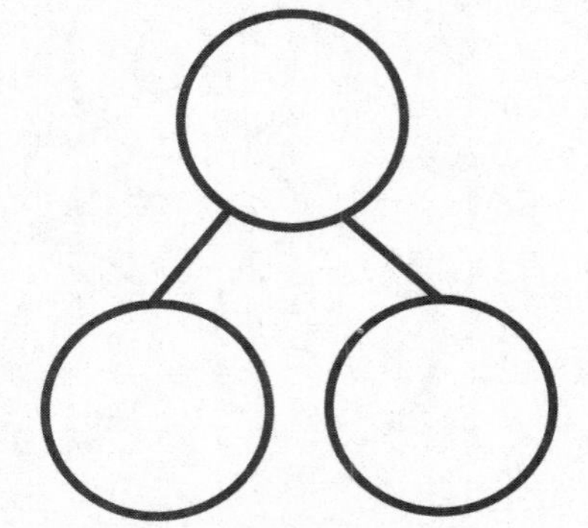

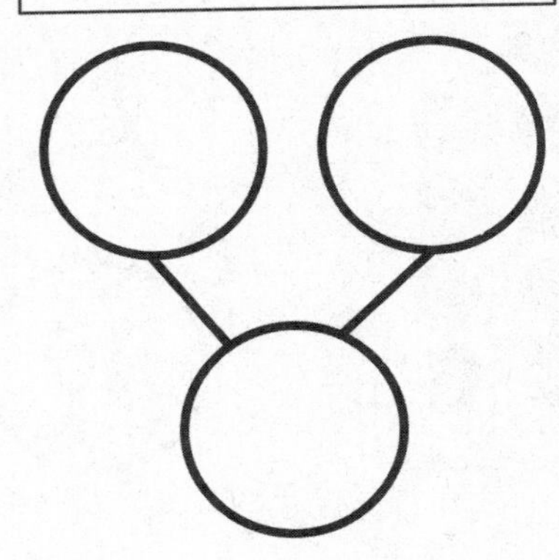

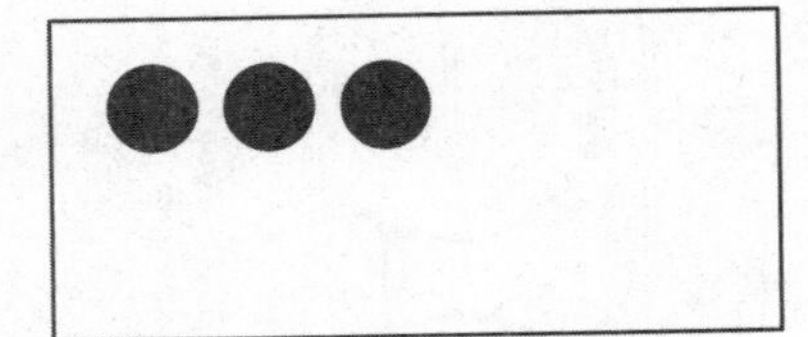

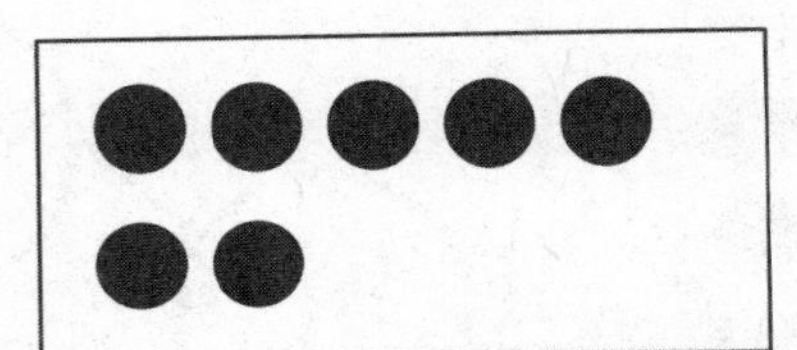

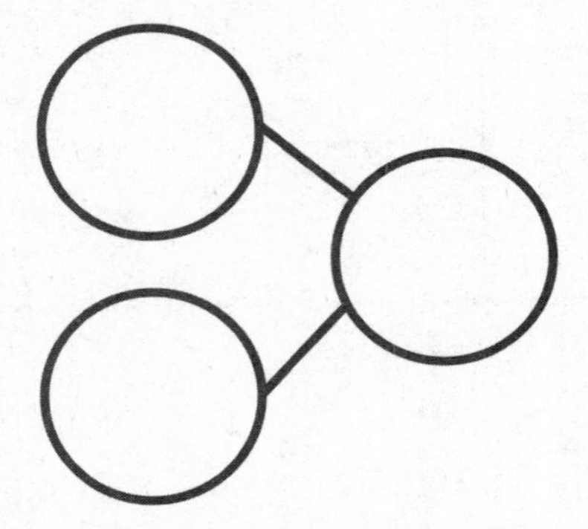

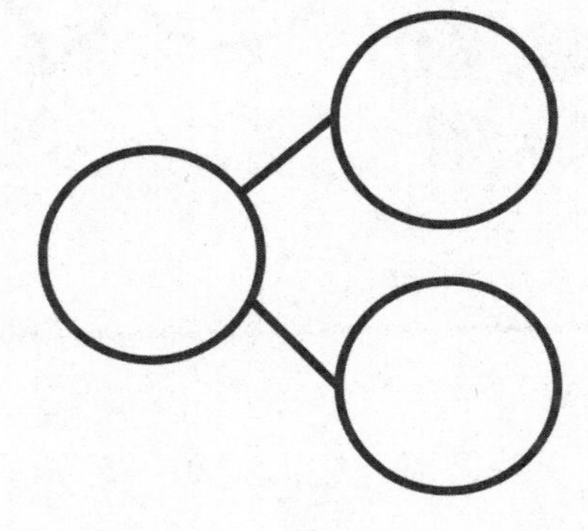

Solve.

9 + 1 = ☐

5 + 5 = ☐

7 + 3 = ☐

10 + 0 = ☐

Ming has 3 baseball caps. There are 10 girls on her team.

Make a 5-group drawing to find how many more caps her team needs.

Draw

Make a number bond about your picture. Share your work with a partner. Do your pictures and number bonds look the same?

_______ + _______ = 10

_______ + _______ = 10

_______ + _______ = 10

_______ + _______ = 10

_______ + _______ = 10

_______ + _______ = 10

_______ + _______ = 10

_______ + _______ = 10

_______ + _______ = 10

add to make 10 recording sheet

Name ______________________________ Date ______________

Look at the 5-group cards. Draw dots to make 10. Fill in the number sentences.

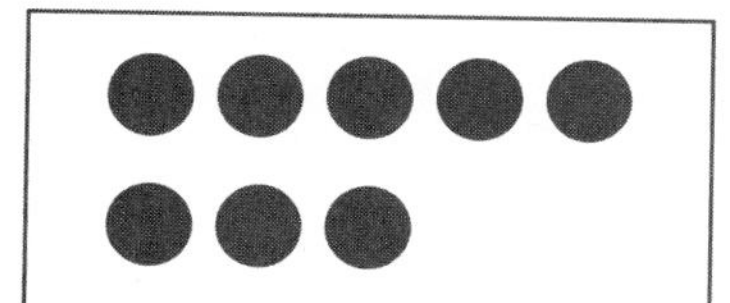

□ + □ = 10

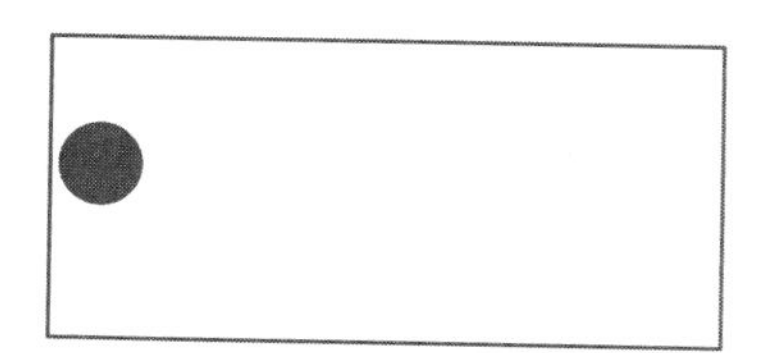

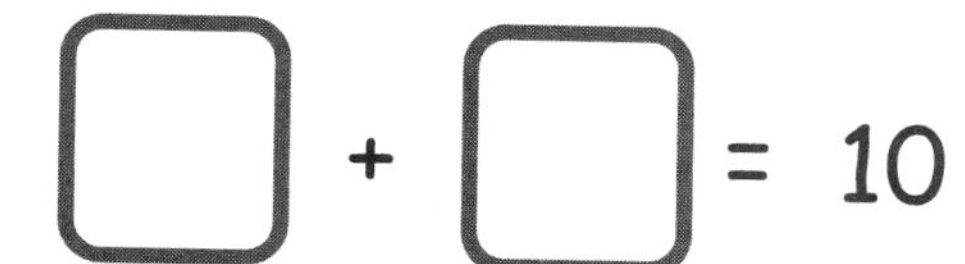

□ + □ = 10

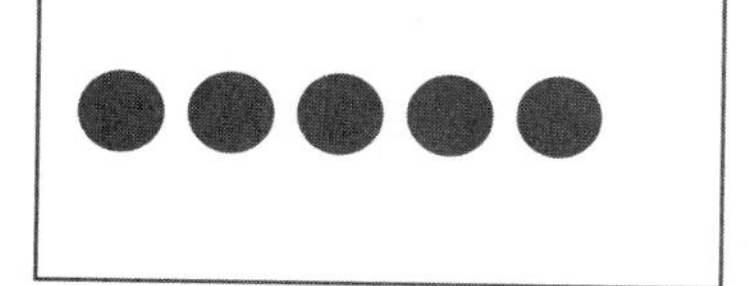

□ + □ = 10

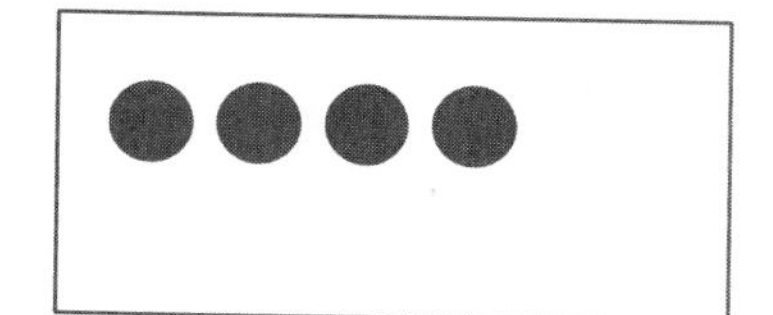

□ + □ = 10

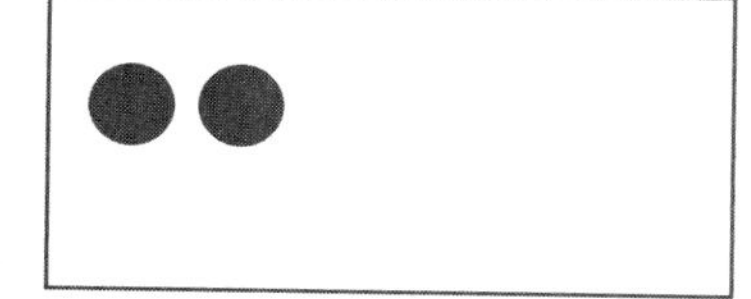

□ + □ = 10

□ + □ = 10

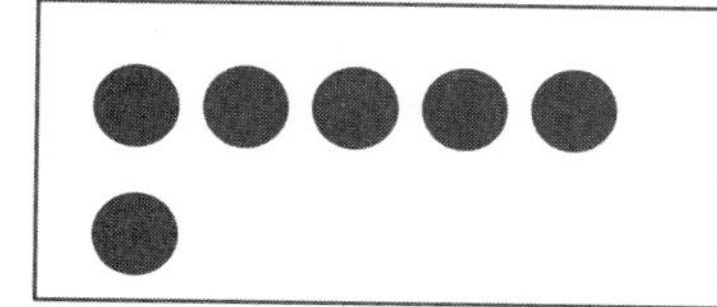

□ + □ = 10

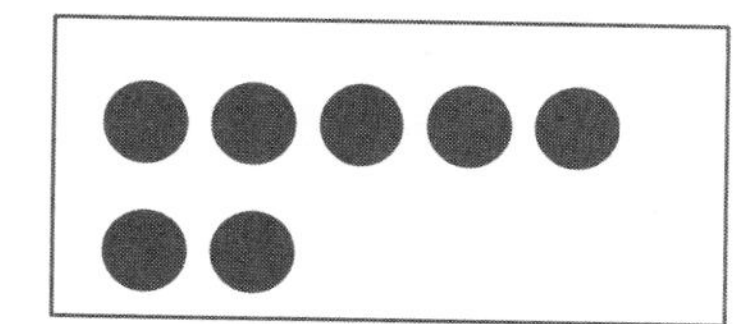

□ + □ = 10

On the back of this page, create a 5-group card. Draw dots to make 10, and write a number sentence.

Credits

Great Minds® has made every effort to obtain permission for the reprinting of all copyrighted material. If any owner of copyrighted material is not acknowledged herein, please contact Great Minds for proper acknowledgment in all future editions and reprints of this module.